BOWLING GREEN STATE UNIVERSITY

DISCARDED

LIBRARY

D1112327

ADAPTIVE IMAGE PROCESSING

A Computational
Intelligence Perspective

IMAGE PROCESSING SERIES

Series Editor: Phillip A. Laplante, Pennsylvania Institute of Technology

Published Titles

Image and Video Compression for Multimedia Engineering
Yun Q. Shi and Huiyang Sun

Multimedia Image and Video Processing
Ling Guan, S.Y. Kung, and Jan Larsen

Shape Analysis and Classification: Theory and Practice
Luciano da Fontoura Costa and Roberto Marcondes Cesar Jr.

Adaptive Image Processing: A Computational Intelligence Perspective
Stuart William Perry, Hau-San Wong, and Ling Guan

Forthcoming Titles

Software Engineering for Image Processing Systems
Phillip A. Laplante

Digital Data-Hiding and Watermarking with Applications
Rajarathnam Chandramouli

ADAPTIVE IMAGE PROCESSING

A Computational Intelligence Perspective

Stuart William Perry
Hau-San Wong
Ling Guan

CRC PRESS

Boca Raton   London   New York   Washington, D.C.

BOWLING GREEN STATE
UNIVERSITY LIBRARIES

Library of Congress Cataloging-in-Publication Data

Perry, Stuart William.
    Adaptive image processing : a computational intelligence perspective / Stuart William Perry, Hau-San Wong, Ling Guan.
        p.   cm. -- (Image processing series)
    Includes bibliographical references and index.
    ISBN 0-8493-0283-8 (alk. paper)
    1. Image processing. 2. Computational intelligence. I. Wong, Hau-San. II. Guan, Ling. III. Title. IV. Series.

TA1637 .P46 2001
006.6--dc21   2001052637

This book contains information obtained from authentic and highly regarded sources. Reprinted material is quoted with permission, and sources are indicated. A wide variety of references are listed. Reasonable efforts have been made to publish reliable data and information, but the author and the publisher cannot assume responsibility for the validity of all materials or for the consequences of their use.

Neither this book nor any part may be reproduced or transmitted in any form or by any means, electronic or mechanical, including photocopying, microfilming, and recording, or by any information storage or retrieval system, without prior permission in writing from the publisher.

The consent of CRC Press LLC does not extend to copying for general distribution, for promotion, for creating new works, or for resale. Specific permission must be obtained in writing from CRC Press LLC for such copying.

Direct all inquiries to CRC Press LLC, 2000 N.W. Corporate Blvd., Boca Raton, Florida 33431.

Trademark Notice: Product or corporate names may be trademarks or registered trademarks, and are used only for identification and explanation, without intent to infringe.

Visit the CRC Press Web site at www.crcpress.com

© 2002 by CRC Press LLC

No claim to original U.S. Government works
International Standard Book Number 0-8493-0283-8
Library of Congress Card Number 2001052637
Printed in the United States of America 1 2 3 4 5 6 7 8 9 0
Printed on acid-free paper

Preface

In this book, we consider the application of computational intelligence techniques to the problem of adaptive image processing. In adaptive image processing, it is usually required to identify each image pixel with a particular feature type (eg., smooth regions, edges, textures, etc.) for separate processing, which constitutes a segmentation problem. We will then establish image models to describe the desired appearance of the respective feature types or, in other words, to characterize each feature type. Finally, we modify the pixel values in such a way that the appearance of the processed features conforms more closely with that specified by the feature models, where the degree of discrepancy is usually measured in terms of a cost function. In other words, we are searching for a set of parameters which minimize this function, i.e., an optimization problem.

To satisfy the above requirements, we consider the application of computational intelligence (CI) techniques to this class of problems. Here we will adopt a specific definition of CI, which includes neural network techniques (NN), fuzzy set theory (FS) and evolutionary computation (EC). A distinguishing characteristic of these algorithms is that they are either biologically inspired, as in the cases of NN and EC, or are attempts to mimic how human beings perceive everyday concepts, as in FS.

The choice of these algorithms is due to the direct correspondence between some of the above requirements with the particular capabilities of specific CI approaches. For example, segmentation can be performed by using NN. In addition, for the purpose of optimization, we can embed the image model parameters as adjustable network weights to be optimized through the network's dynamic action. In contrast, the main role of fuzzy set theory is to address the requirement of characterization, i.e., the specification of human visual preferences, which are usually expressed in fuzzy languages, in the form of multiple fuzzy sets over the domain of pixel value configurations, and the role of EC is mainly in addressing difficult optimization problems.

In this book, we have illustrated the essential aspects of the adaptive image processing problem in terms of two applications: adaptive image restoration and the adaptive characterization of edges in feature detection applications. These two problems are representative of the general adaptive image processing paradigm in that the three requirements of segmentation, characterization and optimization are present.

This book consists of eight chapters. The first chapter provides material of an introductory nature to describe the basic concepts and current state of the art in the field of computational intelligence for image restoration and edge detection. Chapter 2 gives a mathematical description of the restoration problem from the neural network perspective, and describes current algorithms based on this method. Chapter 3 extends the algorithm presented in Chapter 2 to implement adaptive constraint restoration methods for both spatially invariant and spatially variant degradations. Chapter 4 utilizes a perceptually motivated image error measure to introduce novel restoration algorithms. Chapter 5 examines how model-based neural networks can be used to solve image restoration problems. Chapter 6 examines image restoration algorithms making use of the principles of evolutionary computation. Chapter 7 examines the difficult concept of image restoration when insufficient knowledge of the degrading function is available. Finally, Chapter 8 examines the subject of edge detection and characterization using model-based neural networks.

Acknowledgments

We are grateful to our colleagues in the Signal and Multimedia Processing Lab in the University of Sydney, especially Matthew Kyan and Kim Hui Yap, for their contributions and helpful comments during the preparation of this book. Our special thanks to Prof. Terry Caelli for the many stimulating exchanges which eventually led to the work in Chapter 8. We would also like to thank Nora Konopka of CRC Press for her advice and assistance. Finally, we are grateful to our families for their patience and support while we worked on the book.

Contents

Chapter 1

Introduction

1.1 The Importance of Vision

All life-forms require methods for sensing the environment. Being able to sense one's surroundings is of such vital importance for survival that there has been a constant race for life-forms to develop more sophisticated sensory methods through the process of evolution. As a consequence of this process, advanced life-forms have at their disposal an array of highly accurate senses. Some unusual sensory abilities are present in the natural world, such as the ability to detect magnetic and electric fields, or the use of ultrasound waves to determine the structure of surrounding obstacles. Despite this, one of the most prized and universal senses utilized in the natural world is vision.

Advanced animals living aboveground rely heavily on vision. Birds and lizards maximize their fields of view with eyes on each side of their skulls, while other animals direct their eyes forward to observe the world in three dimensions. Nocturnal animals often have large eyes to maximize light intake, while predators such as eagles have very high resolution eyesight to identify prey while flying. The natural world is full of animals of almost every color imaginable. Some animals blend in with surroundings to escape visual detection, while others are brightly colored to attract mates or warn aggressors. Everywhere in the natural world, animals make use of vision for their daily survival. The reason for the heavy reliance on eyesight in the animal world is due to the rich amount of information provided by the visual sense. To survive in the wild, animals must be able to move rapidly. Hearing and smell provide warning regarding the presence of other animals, yet only a small number of animals such as bats have developed these senses sufficiently to effectively utilize the limited amount of information provided by these senses to perform useful actions, such as to escape from predators or chase down prey. For the majority of animals, only vision provides sufficient information in order for them to infer the correct responses under a variety of circumstances.

Humans rely on vision to a much greater extent than most other animals. Unlike the majority of creatures we see in three dimensions with high resolution and color. In humans the senses of smell and hearing have taken second place to vision. Humans have more facial muscles than any other animal, because in our society facial expression is used by each of us as the primary indicator of the emotional states of other humans, rather than the scent signals used by many mammals. In other words, the human world revolves around visual stimuli and the importance of effective visual information processing is paramount for the human visual system.

To interact effectively with the world, the human vision system must be able to extract, process and recognize a large variety of visual structures from the captured images. Specifically, before the transformation of a set of visual stimuli into a meaningful scene, the vision system is required to identify different visual structures such as edges and regions from the captured visual stimuli. Rather than adopting a uniform approach of processing these extracted structures, the vision system should be able to adaptively tune to the specificities of these different structures in order to extract the maximum amount of information for the subsequent recognition stage. For example, the system should selectively enhance the associated attributes of different regions such as color and textures in an adaptive manner such that for some regions, more importance is placed on the extraction and processing of the color attribute, while for other regions the emphasis is placed on the associated textural patterns. Similarly, the vision system should also process the edges in an adaptive manner such that those associated with an object of interest should be distinguished from those associated with the less important ones.

To mimic this adaptive aspect of biological vision and to incorporate this capability into machine vision systems have been the main motivations of image processing and computer vision research for many years. Analogous to the eyes, modern machine vision systems are equipped with one or more cameras to capture light signals, which are then usually stored in the form of digital images or video sequences for subsequent processing. In other words, to fully incorporate the adaptive capabilities of biological vision systems into machines necessitates the design of an effective *adaptive image processing* system. The difficulties of this task can already be foreseen since we are attempting to model a system which is the product of billions of years of evolution and is naturally highly complex. To give machines some of the remarkable capabilities that we take for granted is the subject of intensive ongoing research and the theme of this book.

1.2   Adaptive Image Processing

The need for adaptive image processing arises due to the need to incorporate the above adaptive aspects of biological vision into machine vision systems. For such systems the visual stimuli are usually captured through cameras and presented in the form of digital images which are essentially arrays of pixels, each of which is associated with a gray level value indicating the magnitude of the light signal captured at the corresponding position. To effectively characterize a large variety of image types in image processing, this array of numbers is usually modeled as a 2D discrete non-stationary random process. As opposed to stationary random processes where the statistical properties of the signal remain unchanged with respect to the 2D spatial index, the non-stationary process models the inhomogeneities of visual structures which are inherent in a meaningful visual scene. It is this inhomogeneity that conveys useful information of a scene, usually composed of a number of different objects, to the viewer. On the other hand, a stationary 2D random signal, when viewed as a gray level image, does not usually correspond to the appearances of real-world objects.

For a particular image processing application (we interpret the term "image processing" in a wide sense such that applications in image analysis are also included), we usually assume the existence of an underlying *image model* [1, 2, 3], which is a mathematical description of a hypothetical process through which the current image is generated. If we suppose that an image is adequately described by a stationary random process, which, though not accurate in general, is often invoked as a simplifying assumption, it is apparent that only a single image model corresponding to this random process is required for further image processing. On the other hand, more sophisticated image processing algorithms will account for the non-stationarity of real images by adopting *multiple* image models for more accurate representation. Individual regions in the image can usually be associated with a different image model, and the complete image can be fully characterized by a finite number of these local image models.

1.3 The Three Main Image Feature Classes

The inhomogeneity in images implies the existence of more than one image feature type which convey independent forms of information to the viewer. Although variations among different images can be great, a large number of images can be characterized by a small number of feature types. These are usually summarized under the labels of smooth regions, textures and edges (Figure 1.1). In the following, we will describe the essential characteristics of these three kinds of features, and the image models usually employed for their characterization.

Smooth Regions

Smooth regions usually comprise the largest proportion of areas in images, because surfaces of artificial or natural objects, when imaged from a distance, can usually be regarded as smooth. A simple model for a smooth region is the assignment of a constant gray level value to a restricted domain of the image lattice, together with the addition of Gaussian noise of appropriate variance to model the sensor noise [2, 4].

Edges

As opposed to smooth regions, edges comprise only a very small proportion of areas in images. Nevertheless, most of the information in an image is conveyed

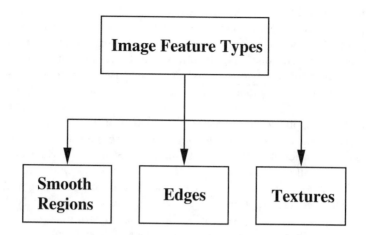

Figure 1.1: The three important classes of feature in images

through these edges. This is easily seen when we look at the edge map of an image after edge detection: we can readily infer the original contents of the image through the edges alone. Since edges represent locations of abrupt transitions of gray level values between adjacent regions, the simplest edge model is therefore a random variable of high variance, as opposed to the smooth region model which uses random variables with low variances. However, this simple model does not take into account the structural constraints in edges, which may then lead to their confusion with textured regions with equally high variances. More sophisticated edge models include the facet model [5], which approximates the different regions of constant gray level values around edges with separate piece-wise continuous functions. There is also the edge profile model, which describes the one-dimensional cross section of an edge in the direction of maximum gray level variation [6, 7]. Attempts have been made to model this profile using a step function and various monotonically increasing functions. Whereas these models mainly characterize the *magnitude* of gray level value transition at the edge location, the edge diagram in terms of zero crossings of the second order gray level derivatives, obtained through the process of Laplacian of Gaussian (LoG) filtering [8, 9], characterizes the edge *positions* in an image. These three edge models are illustrated in Figure 1.2.

Textures

The appearance of textures is usually due to the presence of natural objects in an image. The textures usually have a noise-like appearance, although they are distinctly different from noise in that there usually exists certain discernible patterns within them. This is due to the correlations among the pixel values in specific directions. Due to this noise-like appearance, it is natural to model textures using a 2-D random field. The simplest approach is to use i.i.d (in-

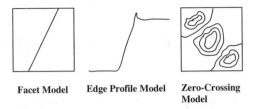

Facet Model   **Edge Profile Model**   **Zero-Crossing Model**

Figure 1.2: Examples of edge models

dependent and identically distributed) random variables with appropriate variances, but this does not take into account the correlations among the pixels. A generalization of this approach is the adoption of Gauss Markov Random Field (GMRF) [10, 11, 12, 13, 14] and Gibbs random field [15, 16] which model these local correlational properties. Another characteristic of textures is their self-similarities: the patterns usually look similar when observed under different magnifications. This leads to their representation as fractal processes [17, 18] which possess this very self-similar property.

1.4   Difficulties in Adaptive Image Processing System Design

Given the very different properties of these three feature types, it is usually necessary to incorporate spatial adaptivity into image processing systems for optimal results. For an image processing system, a set of system parameters is usually defined to control the quality of the processed image. Assuming the adoption of spatial domain processing algorithms, the gray level value x_{i_1,i_2} at spatial index (i_1, i_2) is determined according to the following relationship.

$$x_{i_1,i_2} = f(\mathbf{y}; \mathbf{p}_{SA}(i_1, i_2)) \tag{1.1}$$

In this equation, the mapping f summarizes the operations performed by the image processing system. The vector $\mathbf{y}$ denotes the gray level values of the original image before processing, and $\mathbf{p}_{SA}$ denotes a vector of *spatially adaptive* parameters as a function of the spatial index (i_1, i_2). It is reasonable to expect that different parameter vectors are to be adopted at different positions (i_1, i_2), which usually correspond to different feature types. As a result, an important consideration in the design of this adaptive image processing system is the proper determination of the parameter vector $\mathbf{p}_{SA}(i_1, i_2)$ as a function of the spatial index (i_1, i_2).

On the other hand, for non-adaptive image processing systems, we can simply adopt a constant assignment for $\mathbf{p}_{SA}(i_1, i_2)$

$$\mathbf{p}_{SA}(i_1, i_2) \equiv \mathbf{p}_{NA} \tag{1.2}$$

where $\mathbf{p}_{NA}$ is a constant parameter vector.

We consider examples of $\mathbf{p}_{SA}(i_1, i_2)$ in a number of specific image processing applications below.

- In image filtering, we can define $\mathbf{p}_{SA}(i_1, i_2)$ to be the set of filter coefficients in the convolution mask [2]. Adaptive filtering [19, 20] thus corresponds to using a different mask at different spatial locations, while non-adaptive filtering adopts the same mask for the whole image.

- In image restoration [21, 22, 23], a *regularization parameter* [24, 25, 26] is defined which controls the degree of ill-conditioning of the restoration process, or equivalently, the overall smoothness of the restored image. The vector $\mathbf{p}_{SA}(i_1, i_2)$ in this case corresponds to the scalar regularization parameter. Adaptive regularization [27, 28, 29] involves selecting different parameters at different locations, and non-adaptive regularization adopts a single parameter for the whole image.

- In edge detection, the usual practice is to select a single *threshold* parameter on the gradient magnitude to distinguish between the edge and non-edge points of the image [2, 4], which corresponds to the case of non-adaptive thresholding. This can be considered as a special case of adaptive thresholding, where a threshold value is defined at each spatial location.

Given the above description of adaptive image processing, we can see that the corresponding problem of adaptive parameterization, that of determining the parameter vector $\mathbf{p}_{SA}(i_1, i_2)$ as a function of (i_1, i_2), is particularly acute compared with the non-adaptive case. In the non-adaptive case, and in particular for the case of a parameter vector of low dimensionality, it is usually possible to determine the optimal parameters by interactively choosing different parameter vectors and evaluating the final processed results.

On the other hand, for adaptive image processing, it is almost always the case that a parameter vector of high dimensionality, which consists of the concatenation of all the local parameter vectors, will be involved. If we relax the previous requirement to allow the sub-division of an image into regions and the assignment of the same local parameter vector to each region, the dimension of the resulting concatenated parameter vector can still be large. In addition, the requirement to identify each image pixel with a particular feature type itself constitutes a non-trivial *segmentation* problem. As a result, it is usually not possible to estimate the parameter vector by trial and error. Instead, we should look for a parameter assignment algorithm which would automate the whole process.

To achieve this purpose, we will first have to establish image models which describe the desired local gray level value configurations for the respective image feature types or, in other words, to *characterize* each feature type. Since the local gray level configurations of the processed image are in general a function of the system parameters as specified in equation (1.1), we can associate a *cost function* with each gray level configuration which measures its degree of conformance to the corresponding model, with the local system parameters as arguments of the cost function. We can then search for those system parameter values which minimize the cost function for each feature type, i.e., an *optimization* process. Naturally, we should adopt *different* image models in order to obtain different system parameters for each type of feature.

In view of these requirements, we can summarize the requirements for a successful design of an adaptive image processing system as follows:

Segmentation

Segmentation requires a proper understanding of the difference between the corresponding structural and statistical properties of the various feature types, including those of edges, textures and smooth regions, to allow partition of an image into these basic feature types.

Characterization

Characterization requires an understanding of the most desirable gray level value configurations in terms of the characteristics of the Human Vision System (HVS) for each of the basic feature types, and the subsequent formulation of these criteria into cost functions in terms of the image model parameters, such that the minimization of these cost functions will result in an approximation to the desired gray level configurations for each feature type.

Optimization

In anticipation of the fact that the above criteria will not necessarily lead to well-behaved cost functions, and that some of the functions will be non-linear or even non-differentiable, we should adopt powerful optimization techniques for the searching of the optimal parameter vector.

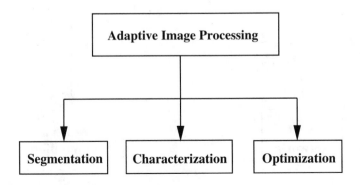

Figure 1.3: The three main requirements in adaptive image processing

These three main requirements are summarized in Figure 1.3.

In this book, our main emphasis is on two specific adaptive image processing systems and their associated algorithms: the adaptive image restoration algorithm and the adaptive edge characterization algorithm. For the former system, segmentation is first applied to partition the image into separate regions according to a local variance measure. Each region then undergoes characterization to establish whether it corresponds to a smooth, edge or textured area. Optimization is then applied as a final step to determine the optimal regularization parameters for each of these regions. For the second system, a preliminary segmentation stage is applied to separate the edge pixels from non-edge pixels. These edge pixels then undergo the characterization process whereby the more salient ones among them (according to the users' preference) are identified. Optimization is finally applied to search for the optimal parameter values for a parametric model of this salient edge set.

1.5   Computational Intelligence Techniques

Considering the above stringent requirements for the satisfactory performance of an adaptive image processing systems, it will be natural to consider the class of algorithms commonly known as computational intelligence techniques. The term "computational intelligence" [30, 31] has sometimes been used to refer to the general attempt to simulate human intelligence on computers, the so-called "artificial intelligence" (AI) approach [32]. However, in this book, we will adopt a more specific definition of computational intelligence techniques which are neural network techniques, fuzzy logic and evolutionary computation (Figure 1.4). These are also referred to as the "numerical" AI approaches (or sometimes "soft

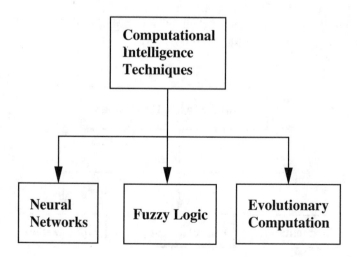

Figure 1.4: The three main classes of computational intelligence algorithms

computing" approach [33]) in contrast to the "symbolic" AI approaches as typified by the expression of human knowledge in terms of linguistic variables in expert systems [32].

A distinguishing characteristic of this class of algorithms is that they are usually biologically inspired: the design of neural networks [34, 35], as the name implies, draws the inspiration mainly from the structure of the human brain. Instead of adopting the serial processing architecture of the Von Neumann computer, a neural network consists of a large number of computational units or neurons (the use of this term again confirming the biological source of inspiration) which are massively interconnected with each other just as the real neurons in the human brain are interconnected with axons and dendrites. Each such connection between the artificial neurons is characterized by an adjustable *weight* which can be modified through a training process such that the overall behavior of the network is changed according to the nature of specific training examples provided, again reminding one of the human learning process.

On the other hand, fuzzy logic [36, 37, 38] is usually regarded as a formal way to describe how human beings perceive everyday concepts: whereas there is no exact height or speed corresponding to concepts like "tall" and "fast," respectively, there is usually a general consensus by humans as to approximately what levels of height and speed the terms are referring to. To mimic this aspect of human cognition on a machine, fuzzy logic avoids the arbitrary assignment of a particular numerical value to a single class. Instead, it defines each such class as a *fuzzy set* as opposed to a *crisp set*, and assigns a fuzzy set membership value within the interval $[0, 1]$ for each class which expresses the degree of membership of the particular numerical value in the class, thus generalizing the previous concept of crisp set membership values within the discrete set $\{0, 1\}$.

For the third member of the class of computational intelligence algorithms, no concept is closer to biology than the concept of evolution, which is the incremental adaptation process by which living organisms increase their fitness to survive in a hostile environment through the processes of mutation and competition. Central to the process of evolution is the concept of a *population* in which the better adapted individuals gradually displace the not so well adapted ones. Described within the context of an optimization algorithm, an *evolutionary computational algorithm* [39, 40] mimics this aspect of evolution by generating a population of potential solutions to the optimization problem, instead of a sequence of single potential solution as in the case of gradient descent optimization or simulated annealing [16]. The potential solutions are allowed to compete against each other by comparing their respective cost function values associated with the optimization problem with each other. Solutions with high cost function values are displaced from the population while those with low cost values survive into the next generation. The displaced individuals in the population are replaced by generating new individuals from the survived solutions through the processes of mutation and recombination. In this way, many regions in the search space can be explored simultaneously, and the search process is not affected by local minima as no gradient evaluation is required for this algorithm.

We will now have a look at how the specific capabilities of these computational intelligence techniques can address the various problems encountered in the design and parameterization of an adaptive image processing system.

Neural Networks

Artificial neural network represents one of the first attempts to incorporate learning capabilities into computing machines. Corresponding to the biological neurons in human brain, we define artificial neurons which perform simple mathematical operations. These artificial neurons are connected with each other through *network weights* which specify the strength of the connection. Analogous to its biological counterpart, these network weights are adjustable through a learning process which enables the network to perform a variety of computational tasks. The neurons are usually arranged in *layers*, with the input layer accepting signals from the external environment, and the output layer emitting the result of the computations. Between these two layers are usually a number of *hidden* layers which perform the intermediate steps of computations. The architecture of a typical artificial neural network with one hidden layer is shown in Figure 1.5. In specific types of network, the hidden layers may be missing and only the input and output layers are present.

The adaptive capability of neural networks through the adjustment of the network weights will prove useful in addressing the requirements of segmentation, characterization and optimization in adaptive image processing system design. For segmentation, we can, for example, ask human users to specify which part of an image corresponds to edges, textures and smooth regions, etc. We can

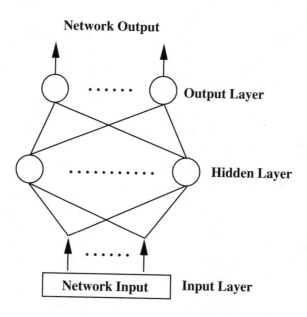

Figure 1.5: The architecture of a neural network with one hidden layer

then extract image features from the specified regions as training examples for a properly designed neural network such that the trained network will be capable of segmenting a previously unseen image into the primitive feature types. Previous works where neural network is applied to the problem of image segmentation are detailed in [41, 42, 43].

Neural network is also capable of performing characterization to a certain extent, especially in the process of *unsupervised competitive learning* [34, 44], where both segmentation and characterization of training data are carried out: during the competitive learning process, individual neurons in the network, which represent distinct sub-classes of training data, gradually build up *templates* of their associated sub-classes in the form of *weight vectors*. These templates serve to characterize the individual sub-classes.

In anticipation of the possible presence of non-linearity in the cost functions for parameter estimation during the optimization process, neural network is again an ideal candidate for accommodating such difficulties: the operation of a neural network is inherently non-linear due to the presence of the sigmoid neuronal transfer function. We can also tailor the non-linear neuronal transfer function specifically to a particular application. More generally, we can *map* a cost function onto a neural network by adopting an architecture such that the image model parameters will appear as adjustable weights in the network [45, 46]. We can then search for the optimal image model parameters by minimizing the embedded cost function through the dynamic action of the neural network.

In addition, while the distributed nature of information storage in neural networks and the resulting fault-tolerance is usually regarded as an overriding factor in its adoption, we will, in this book, concentrate rather on the possibility of task localization in a neural network: we will sub-divide the neurons into *neuron clusters*, with each cluster specialized for the performance of a certain task [47, 48]. It is well known that similar localization of processing occurs in the human brain, as in the classification of the cerebral cortex into visual area, auditory area, speech area and motor area, etc. [49, 50]. In the context of adaptive image processing, we can, for example, sub-divide the set of neurons in such a way that each cluster will process the three primitive feature types, namely, textures, edges and smooth regions, respectively. The values of the connection weights in each sub-network can be different, and we can even adopt different architectures and learning strategies for each sub-network for optimal processing of its assigned feature type.

Fuzzy Logic

From the previous description of fuzzy techniques, it is obvious that its main application in adaptive image processing will be to address the requirement of characterization, i.e., the specification of human visual preferences in terms of gray level value configurations. Many concepts associated with image processing are inherently fuzzy, such as the description of a region as "dark" or "bright," and the incorporation of fuzzy set theory is usually required for satisfactory processing results [51, 52, 53, 54, 55]. The very use of the words "textures," "edges" and "smooth regions" to characterize the basic image feature types implies fuzziness: the difference between smooth regions and weak textures can be subtle, and the boundary between textures and edges is sometimes blurred if the textural patterns are strongly correlated in a certain direction so that we can regard the pattern as multiple edges. Since the image processing system only recognizes gray level configurations, it will be natural to define fuzzy sets with qualifying terms like "texture," "edge" and "smooth regions" over the set of corresponding gray level configurations according to human preferences. However, one of the problems with this approach is that there is usually an extremely large number of possible gray level configurations corresponding to each feature type, and human beings cannot usually relate what they perceive as a certain feature type to a particular configuration. In Chapter 5, a *scalar* measure has been established which characterizes the degree of resemblance of a gray level configuration to either textures or edges. In addition, we can establish the exact interval of values of this measure where the configuration will more resemble textures than edges, and *vice versa*. As a result, we can readily define fuzzy sets over this one-dimensional *universe of discourse* [37].

In addition, fuzzy set theory also plays an important role in the derivation of improved segmentation algorithms. A notable example is the *fuzzy c-means algorithm* [56, 57, 58, 59], which is a generalization of the *k-means algorithm* [60] for data clustering. In the *k*-means algorithm, each data vector, which may contain feature values or gray level values as individual components in image processing applications, is assumed to belong to one and only one class. This may result in inadequate characterization of certain data vectors which possess properties common to more than one class, but then get arbitrarily assigned to one of those classes. This is prevented in the fuzzy *c*-means algorithm, where each data vector is assumed to belong to every class to a different degree which is expressed by a numerical membership value in the interval $[0, 1]$. This paradigm can now accommodate those data vectors which possess attributes common to more than one class, in the form of large membership values in several of these classes.

Evolutionary Computation

The often stated advantages of evolutionary computation include its implicit parallelism which allows simultaneous exploration of different regions of the search space [61], and its ability to avoid local minima [39, 40]. However, in this book, we will emphasize its capability to search for the optimizer of a *non-differentiable* cost function efficiently, i.e., to satisfy the requirement of optimization. An example of a non-differentiable cost function in image processing would be the metric which compares the probability density function (pdf) of a certain local attribute of the image (gray level values, gradient magnitudes, etc.) with a desired pdf. We would, in general, like to adjust the parameters of the adaptive image processing system in such a way that the distance between the pdf of the processed image is as close as possible to the desired pdf. In other words, we would like to minimize the distance as a function of the system parameters. In practice, we

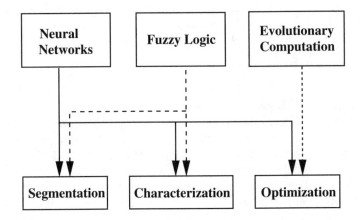

Figure 1.6: Relationships between the computational intelligence algorithms and the main requirements in adaptive image processing

have to approximate the pdfs using histograms of the corresponding attributes, which involves the counting of discrete quantities. As a result, although the pdf of the processed image is a function of the system parameters, it is not differentiable with respect to these parameters. Although stochastic algorithms like simulated annealing can also be applied to minimize non-differentiable cost functions, evolutionary computational algorithms represent a more efficient optimization approach due to the implicit parallelism of its population-based search strategy.

The relationship between the main classes of algorithms in computational intelligence and the major requirements in adaptive image processing is summarized in Figure 1.6.

1.6 Scope of the Book

In this book, as specific examples of adaptive image processing systems, we consider the *adaptive regularization* problem in image restoration [27, 28, 29] and the edge characterization problem in image analysis. We adopt the neural network technique as our main approach to these problems due to its capability to satisfy all three requirements in adaptive image processing, as illustrated in Figure 1.6. In particular, we use a specialized form of network known as *model-based neural network with hierarchical architecture* [48, 62]. The reason for its adoption is that its specific architecture, which consists of a number of model-based *sub-networks*, particularly facilitates the implementation of adaptive image processing applications, where each sub-network can be specialized to process a particular type of image features. In addition to NN, fuzzy techniques and evolutionary computational algorithms, the other two main techniques of computational intelligence, are adopted as complementary approaches for the adaptive image processing problem, especially in view of its associated requirements of characterization and optimization as described previously.

1.6.1   Image Restoration

The act of attempting to obtain the original image given the degraded image and some knowledge of the degrading factors is known as *image restoration*. The problem of restoring an original image, when given the degraded image, with or without knowledge of the degrading *point spread function* (PSF) or degree and type of noise present is an ill-posed problem [21, 24, 63, 64] and can be approached in a number of ways such as those given in [21, 65, 66, 67]. For all useful cases a set of simultaneous equations is produced which is too large to be solved analytically. Common approaches to this problem can be divided into two categories, inverse filtering or transform related techniques, and algebraic techniques. An excellent review of classical image restoration techniques is given by [21]. The following references also contain surveys of restoration techniques, Katsaggelos [23], Sondhi [68], Andrews [69], Hunt [70], and Frieden [71].

Image Degradations

Since our imaging technology is not perfect, every recorded image is a degraded image in some sense. Every imaging system has a limit to its available resolution and the speed at which images can be recorded. Often the problems of finite resolution and speed are not crucial to the applications of the images produced, but there are always cases where this is not so. There exists a large number of possible degradations that an image can suffer. Common degradations are blurring, motion and noise. Blurring can be caused when an object in the image is outside the cameras depth of field some time during the exposure. For example, a foreground tree might be blurred when we have set up a camera with a telephoto lens to take a photograph of a mountain. A blurred object loses some small scale detail and the blurring process can be modeled as if high frequency components have been attenuated in some manner in the image [4, 21]. If an imaging system internally attenuates the high frequency components in the image, the result will again appear blurry, despite the fact that all objects in the image were in the camera's field of view. Another commonly encountered image degradation is motion blur. Motion blur can be caused when a object moves relative to the camera during an exposure, such as a car driving along a highway in an image. In the resultant image, the object appears to be smeared in one direction. Motion blur can also result when the camera moves during the exposure. Noise is generally a distortion due to the imaging system rather than the scene recorded. Noise results in random variations to pixel values in the image. This could be caused by the imaging system itself, or the recording or transmission medium. Sometimes the definitions are not clear as in the case where an image is distorted by atmospheric turbulence, such as heat haze. In this case, the image appears blurry because the atmospheric distortion has caused sections of the object to be imaged to move about randomly. This distortion could be described as random motion blur, but can often be modeled as a standard blurring process. Some types of image distortions, such as certain types of atmospheric degradations [72, 73, 74, 75, 76], can be best described as distortions in the phase of the signal. Whatever the degrading process, image distortions can fall into two categories [4, 21].

- Some distortions may be described as *spatially invariant* or *space invariant*. In a space invariant distortion all pixels have suffered the same form of distortion. This is generally caused by problems with the imaging system such as distortions in optical system, global lack of focus or camera motion.

- General distortions are what is called spatially variant or space variant. In a space variant distortion, the degradation suffered by a pixel in the image depends upon its location in the image. This can be caused by internal factors, such as distortions in the optical system, or by external factors, such as object motion.

In addition, image degradations can be described as linear or non-linear [21]. In this book, we consider only those distortions which may be described by a linear model. For these distortions, a suitable mathematical model is given in Chapter 2.

Adaptive Regularization

In regularized image restoration, the associated cost function consists of two terms: a data conformance term which is a function of the degraded image pixel values and the degradation mechanism, and the model conformance term which is usually specified as a continuity constraint on neighboring gray level values to alleviate the problem of ill-conditioning characteristic of this kind of inverse problems. The *regularization parameter* [23, 25] controls the relative contributions of the two terms toward the overall cost function.

In general, if the regularization parameter is increased, the model conformance term is emphasized at the expense of the data conformance term, and the restored image becomes smoother while the edges and textured regions become blurred. On the contrary, if we decrease the parameter, the fidelity of the restored image is increased at the expense of decreased noise smoothing. If a single parameter value is used for the whole image, it should be chosen such that the quality of the resulting restored image would be a compromise between the above two extremes.

More generally, we can adopt different regularization parameter values for regions in the image corresponding to different feature types. This is more desirable due to the different noise masking capabilities of distinct feature types: since noise is more visible in the smooth regions, we should adopt a larger parameter value in those regions, while we could use a smaller value in the edge and textured regions to enhance the details there due to their greater noise masking capabilities. We can even further distinguish between the edge and textured regions and assign a still smaller parameter value to the textured regions due to their closer resemblance to noises.

Adaptive regularization can thus be regarded as a representative example of the design of an adaptive image processing system, since the stages of segmentation, characterization and optimization are included in its implementation: the segmentation stage consists of the partitioning of the image into its constituent feature types, the characterization stage involves specifying the desired gray level configurations for each feature type after restoration, and relating these configurations to particular values of the regularization parameter in terms of various image models and the associated cost functions. The final optimization stage searches for the optimal parameter values by minimizing the resulting cost functions. Since the hierarchical model-based neural networks can satisfy each of the above three requirements to a certain extent, we propose using such networks to solve this problem as a first step.

The selection of this particular image processing problem is by no means restrictive, as the current framework can be generalized to a large variety of related processing problems: in adaptive image enhancement [77, 78], it is also desirable to adopt different enhancement criteria for different feature types. In adaptive image filtering, we can derive different sets of filter coefficients for the convolution mask in such a way that the image details in the edge and textured regions are preserved, while the noise in the smooth regions are attenuated. In segmentation-based image compression [79, 80, 81], the partitioning of the image into its constituent features are also required in order to assign different statistical models and their associated optimal quantizers to the respective features.

Perception-Based Error Measure for Image Restoration

The most common method to compare the similarity of two images is to compute their mean square error (MSE). However, the MSE relates to the power of the error signal and has little relationship to human visual perception. An important drawback to the MSE and any cost function which attempts to use the MSE to restore a degraded image is that the MSE treats the image as a stationary process. All pixels are given equal priority regardless of their relevance to human perception. This suggests that information is ignored. When restoring images for the purpose of better clarity as perceived by humans the problem becomes acute. When humans observe the differences between two images, they do not give much consideration to the differences in individual pixel level values. Instead humans are concerned with matching edges, regions and textures between the two images. This is contrary to the concepts involved in the MSE. From this it can be seen that any cost function which treats an image as a stationary process can only produce a sub-optimal result. In addition, treating the image as a stationary process is contrary to the principles of computational intelligence.

Humans tend to pay more attention to sharp differences in intensity within an image [82, 83, 84], for example, edges or noise in background regions. Hence an error measure should take into account the concept that low variance regions in the original image should remain low variance regions in the enhanced image, and high variance regions should likewise remain high variance regions. This implies that noise should be kept at a minimum in background regions, where it is most noticeable, but noise suppression should not be as important in highly textured regions where image sharpness should be the dominant consideration. These considerations are especially important in the field of color image restoration. Humans appear to be much more sensitive to slight color variations than they are to variations in brightness.

Considerations regarding human perception have been examined in the past [82, 83, 85, 86, 87, 88, 89, 90, 91, 92, 93]. A great deal of work has been done toward developing linear filters for the removal of noise which incorporate some model of human perception [84, 94, 95]. In these works it is found that edges have a great importance to the way humans perceive images. Ran and Farvardin considered psychovisual properties of the human visual system in order to develop a technique to decompose an image into smooth regions, textured regions and regions containing what are described as *strong edges* [96]. This was done with a view primarily toward image compression. Similarly, Bellini, Leone and Rovatti developed a fuzzy perceptual classifier to create what they described as a *pixel relevance map* to aid in image compression [97]. Hontsch and Karam developed a perceptual model for image compression which decomposed the image into components with varying frequency and orientation [98]. A perceptual distortion measure was then described which used a number of experimentally derived constants. Huang and Coyle considered the use of stack filters for image restoration [99]. They used the concept of a weighted mean absolute error (WMAE), where the weights were determined by the perceptually motivated *visible differences predictor* (VDP) described in [100]. In the past, research has for the most part been concerned with the preservation of edges and the reduction of ringing effects caused by low-pass filters and the models presented to take account of human perception are often complicated. However, in this book we will show that simple functions which incorporate some psychovisual properties of the human visual system can be easily incorporated into existing algorithms and can provide improvements over current techniques.

In view of the problems with classical error measures such as the MSE, Perry and Guan [101] and Perry [102] presented a different error measure, *local standard deviation mean square error* (LSMSE), which is based on the comparison of local standard deviations in the neighborhood of each pixel instead of their gray level values. The LSMSE is calculated in the following way: Each pixel in the two images to be compared has its local standard deviation calculated over a small neighborhood centered on the pixel. The error between each pixel's local standard deviation in the first image and the corresponding pixel's local standard deviation in the second image is computed. The LSMSE is the mean squared error of these differences over all pixels in the image. The mean square error between the two standard deviations gives an indication of the degree of similarity between the two images. This error measure requires matching between the high and low variance regions of the image, which is more intuitive in terms of human visual perception.

This alternative error measure will be heavily relied upon in Chapters 2,

3 and 4 and is hence presented here. A mathematical description is given in Chapter 4.

Blind Deconvolution

In comparison with the determination of the regularization parameter for image restoration, the problem of *blind deconvolution* is considerably more difficult, since in this case the degradation mechanism, or equivalently the form of the point spread function, is unknown to the user. As a result, in addition to estimating the local regularization parameters, we have to estimate the coefficients of the point spread function itself. In Chapter 7, we describe an approach for blind deconvolution which is based on computational intelligence techniques. Specifically, the blind deconvolution problem is first formulated within the framework of evolutionary strategy where a pool of candidate PSFs are generated to form the population in ES. A new cost function which incorporates the specific requirement of blind deconvolution in the form of a point spread function domain regularization term, which ensures the emergence of a valid PSF, in addition to the previous data fidelity measure and image regularization term is adopted as the fitness function in the evolutionary algorithm. This new blind deconvolution approach will be described in Chapter 7.

1.6.2   Edge Characterization and Detection

The characterization of important features in an image requires the detailed specification of those pixel configurations which human beings would regard as significant. In this work, we consider the problem of representing human preferences, especially with regard to image interpretation, again in the form of a model-based neural network with hierarchical architecture [48, 62, 103]. Since it is difficult to represent all aspects of human preferences in interpreting images using traditional mathematical models, we encode these preferences through a direct *learning* process, using image pixel configurations which humans usually regard as visually significant as training examples. As a first step, we consider the problem of edge characterization in such a network. This representation problem is important since its successful solution would allow computer vision systems to simulate to a certain extent the decision process of human beings when interpreting images.

Whereas the network can be considered as a particular implementation of the stages of segmentation and characterization in the overall adaptive image processing scheme, it can also be regarded as a self-contained adaptive image processing system on its own: the network is designed such that it automatically partitions the edges in an image into different classes depending on the gray level values of the surrounding pixels of the edge, and applies different detection thresholds to each of the classes. This is in contrast to the usual approach where a single detection threshold is adopted across the whole image independent of the local context. More importantly, instead of providing *quantitative* values for the threshold as in the usual case, the users are asked to provide *qualitative* opinions on what they regard as edges by manually tracing their desired edges on an image. The gray level configurations around the trace are then used as training examples for the model-based neural network to acquire an internal model of the edges, which is another example of the design of an adaptive image processing system through the training process.

As seen above, we have proposed the use of hierarchical model-based neural network for the solution of both these problems as a first attempt. It was observed later that, whereas the edge characterization problem can be satisfactorily represented by this framework, resulting in adequate characterization of those image edges which humans regard as significant, there are some inadequacies in using this framework exclusively for the solution of the adaptive regularization problem, especially in those cases where the images are more severely degraded. These inadequacies motivate our later adoption of fuzzy set theory and evolutionary computation techniques, in addition to the previous neural network techniques, for this problem.

1.7 Contributions of the Current Work

With regard to the problems posed by the requirements of segmentation, characterization and optimization in the design of an adaptive image processing system, we have devised a system of interrelated solutions comprising the use of the main algorithm classes of computational intelligence techniques. The contributions of the work described in this book can be summarized as follows.

1.7.1 Application of Neural Networks for Image Restoration

Different neural network models, which will be described in Chapters 2, 3, 4 and 5, are adopted for the problem of image restoration. In particular, a model-based neural network with hierarchical architecture [48, 62, 103] is derived for the problem of adaptive regularization. The image is segmented into smooth regions and combined edge/textured regions, and we assign a single sub-network to each of these regions for the estimation of the regional parameters. An important new concept arising from this work is our alternative viewpoint of the regularization parameters as model-based neuronal weights, which are then trainable through the supply of proper training examples. We derive the training examples through the application of *adaptive non-linear filtering* [104] to individual pixel neighborhoods in the image for an *independent* estimate of the current pixel value.

1.7.2   Application of Neural Networks to Edge Characterization

A model-based neural network with hierarchical architecture is proposed for the problem of edge characterization and detection. Unlike previous edge detection algorithms where various threshold parameters have to be specified [2, 4], this parameterization task can be performed implicitly in a neural network by supplying training examples. The most important concept in this part of the work is to allow human users to communicate their preferences to the adaptive image processing system through the provision of qualitative training examples in the form of edge tracings on an image, which is a more natural way of specifying preferences for humans, than the selection of *quantitative* values for a set of parameters. With the adoption of this network architecture and the associated training algorithm, it will be shown that the network can generalize from sparse examples of edges provided by human users to detect all significant edges in images not in the training set. More importantly, no re-training and alteration of architecture is required for applying the same network to noisy images, unlike conventional edge detectors which usually require threshold re-adjustment.

1.7.3   Application of Fuzzy Set Theory to Adaptive Regularization

For the adaptive regularization problem in image restoration, apart from the requirement of adopting different regularization parameters for smooth regions and regions with high gray level variances, it is also desirable to further separate the latter regions into edge and textured regions. This is due to the different noise masking capabilities of these two feature types, which in turn requires different regularization parameter values. In our previous discussion of fuzzy set theory, we have described a possible solution to this problem, in the form of characterizing the gray level configurations corresponding to the above two feature types, and then define fuzzy sets with qualifying terms like "texture" and "edge" over the respective sets of configurations. However, one of the problems with this approach is that there is usually an extremely large number of possible gray level configurations corresponding to each feature type, and human beings cannot usually relate what they perceive as a certain feature type to a particular configuration. In Chapter 5, a *scalar* measure has been established which characterizes the degree of resemblance of a gray level configuration to either textures or edges. In addition, we can establish the exact interval of values of this measure where the configuration will more resemble textures than edges, and *vice versa*. As a result, we can readily define fuzzy sets over this one-dimensional *universe of discourse* [37].

1.7.4 Application of Evolutionary Programming to Adaptive Regularization and Blind Deconvolution

Apart from the neural network-based techniques, we have developed an alternative solution to the problem of adaptive regularization using evolutionary programming, which is a member of the class of evolutionary computational algorithms [39, 40]. Returning again to the ETC measure, we have observed that the distribution of the values of this quantity assumes a typical form for a large class of images. In other words, the shape of the probability density function (pdf) of this measure is similar across a broad class of images and can be modeled using piecewise continuous functions. On the other hand, this pdf will be different for blurred images or incorrectly regularized images. As a result, the model pdf of the ETC measure serves as a kind of *signature* for correctly regularized images, and we should minimize the difference between the corresponding pdf of the image being restored and the model pdf using some kind of distance measure. The requirement to approximate this pdf using a histogram, which involves the counting of discrete quantities, and the resulting non-differentiability of the distance measure with respect to the various regularization parameters, necessitates the use of evolutionary computational algorithms for optimization. We have adopted evolutionary programming that, unlike the genetic algorithm which is another widely applied member of this class of algorithms, operates directly on real-valued vectors instead of binary-coded strings and is therefore more suited to the adaptation of the regularization parameters. In this algorithm, we have derived a parametric representation which expresses the regularization parameter value as a function of the local image variance. Generating a population of these *regularization strategies* which are vectors of the above hyperparameters, we apply the processes of mutation, competition and selection to the members of the population to obtain the optimal regularization strategy. This approach is then further extended to solve the problem of blind deconvolution by including the point spread function coefficients in the set of hyperparameters associated with each individual in the population.

1.8 Overview of This Book

This book consists of eight chapters. The first chapter provides material of an introductory nature to describe the basic concepts and current state of the art in the field of computational intelligence for image restoration and edge detection. Chapter 2 gives a mathematical description of the restoration problem from the Hopfield neural network perspective, and describes current algorithms based on this method. Chapter 3 extends the algorithm presented in Chapter 2 to implement adaptive constraint restoration methods for both spatially invariant and spatially variant degradations. Chapter 4 utilizes a perceptually motivated image error measure to introduce novel restoration algorithms. Chapter 5 examines how model-based neural networks [62] can be used to solve image restoration problems. Chapter 6 examines image restoration algorithms making use of the principles of evolutionary computation. Chapter 7 examines the difficult concept of image restoration when insufficient knowledge of the degrading function is available. Finally, Chapter 8 examines the subject of edge detection and characterization using model-based neural networks.

Chapter 2

Fundamentals of Neural Network Image Restoration

2.1 Image Distortions

Images are often recorded under a wide variety of circumstances. As imaging technology is rapidly advancing, our interest in recording unusual or irreproducible phenomena is increasing as well. We often push imaging technology to its very limits. For this reason we will always have to handle images suffering from some form of degradation.

Since our imaging technology is not perfect, every recorded image is a degraded version of the scene in some sense. Every imaging system has a limit to its available resolution and the speed at which images can be recorded. Often the problems of finite resolution and speed are not crucial to the applications of the images produced, but there are always cases where this is not so. There exists a large number of possible degradations that an image can suffer. Common degradations are blurring, motion and noise. Blurring can be caused when an object in the image is outside the cameras depth of field some time during the exposure. For example, a foreground tree might be blurred when we have set up a camera with a telephoto lens to take a photograph of a distant mountain. A blurred object loses some small scale detail and the blurring process can be modeled as if high frequency components have been attenuated in some manner in the image [4, 21]. If an imaging system internally attenuates the high frequency components in the image, the result will again appear blurry, despite the fact that all objects in the image were in the camera's field of view. Another commonly encountered image degradation is motion blur. Motion blur can be caused when a object moves relative to the camera during an exposure, such as a car driving along a highway in an image. In the resultant image, the object appears to be smeared in one direction. Motion blur can also result when the camera moves during the exposure. Noise is generally a distortion due to the imaging system rather than the scene recorded. Noise results in random variations to pixel values in the image. This could be caused by the imaging system itself, or the recording or transmission medium. Sometimes the definitions are not clear as in the case where an image is distorted by atmospheric turbulence, such as heat haze. In this case, the image appears blurry because the atmospheric distortion has caused sections of the object to be imaged to move about randomly. This distortion could be described as random motion blur, but can often be modeled as a standard blurring process. Some types of image distortions, such as certain types of atmospheric degradations [72, 73, 74, 75, 76], can be best described as distortions in the phase of the signal. Whatever the degrading process, image distortions may be placed into two categories [4, 21].

- Some distortions may be described as spatially invariant or space invariant. In a space invariant distortion, the parameters of the distortion function are kept unchanged for all regions of the image and all pixels suffer the same form of distortion. This is generally caused by problems with the imaging system such as distortions in the optical system, global lack of focus or camera motion.

- General distortions are what is called spatially variant or space variant. In a space variant distortion, the degradation suffered by a pixel in the image depends upon its location in the image. This can be caused by internal factors, such as distortions in the optical system, or by external factors, such as object motion.

In addition, image degradations can be described as linear or non-linear [21]. In this book, we consider only those distortions which may be described by a linear model.

All linear image degradations can be described by their impulse response. A two-dimensional impulse response is often called a Point Spread Function (PSF). It is a two-dimensional function that smears a pixel at its center with some of the pixel's neighbors. The size and shape of the neighborhood used by the PSF is called the PSF's *Region of Support*. Unless explicitly stated, we will from now on consider PSFs with square shaped neighborhoods. The larger the neighborhood, the more smearing occurs and the worse the degradation to the image. Here is an example of a 3 by 3 discrete PSF.

$$\frac{1}{5} \begin{bmatrix} 0.5 & 0.5 & 0.5 \\ 0.5 & 1.0 & 0.5 \\ 0.5 & 0.5 & 0.5 \end{bmatrix}$$

where the factor $\frac{1}{5}$ ensures energy conservation.

The final value of the pixel acted upon by this PSF is the sum of the values of each pixel under the PSF mask, each multiplied by the matching entry in the PSF mask.

Consider a PSF of size P by P acting on an image of size N by M. In the case of a two-dimensional image, the PSF may be written as $h(x, y; \alpha, \beta)$. The four sets of indices indicate that the PSF may be spatially variant hence the PSF will be a different function for pixels in different locations of an image. When noise is also present in the degraded image, as is often the case in real-world applications, the image degradation model in the discrete case becomes [4]:

$$g(x, y) = \sum_{\alpha}^{N} \sum_{\beta}^{M} f(\alpha, \beta) h(x, y; \alpha, \beta) + n(x, y) \tag{2.1}$$

where $f(x, y)$ and $g(x, y)$ are the original and degraded images, respectively, and $n(x, y)$ is the additive noise component of the degraded image. If $h(x, y; \alpha, \beta)$ is a linear function then (2.1) may be restated by lexicographically ordering $g(x, y)$, $f(x, y)$ and $n(x, y)$ into column vectors of size NM. To lexicographically order an image, we simply scan each pixel in the image row by row and stack them one after another to form a single column vector. Alternately, we may scan the image column by column to form the vector. For example, assume the image $f(x, y)$ looks like:

$$f(x, y) = \begin{bmatrix} 11 & 12 & 13 & 14 \\ 21 & 22 & 23 & 24 \\ 31 & 32 & 33 & 34 \\ 41 & 42 & 43 & 44 \end{bmatrix}$$

After lexicographic ordering the following column vector results:

$$\mathbf{f} = [11\ 12\ 13\ 14\ 21\ 22\ 23\ 24\ 31\ 32\ 33\ 34\ 41\ 42\ 43\ 44]^T$$

If we are consistent and order $g(x, y)$, $f(x, y)$ and $n(x, y)$ and in the same way, we may restate (2.1) as a matrix operation [4, 21]:

$$\mathbf{g} = \mathbf{H}\mathbf{f} + \mathbf{n} \tag{2.2}$$

where $\mathbf{g}$ and $\mathbf{f}$ are the lexicographically organized degraded and original image vectors, $\mathbf{n}$ is the additive noise component and $\mathbf{H}$ is a matrix operator whose elements are an arrangement of the elements of $h(x, y; \alpha, \beta)$ such that the matrix multiplication of $\mathbf{f}$ with $\mathbf{H}$ performs the same operation as convolving $f(x, y)$ with $h(x, y; \alpha, \beta)$. In general, $\mathbf{H}$ may take any form. However, if $h(x, y; \alpha, \beta)$ is spatially invariant with $P \ll \min(N, M)$ then $h(x, y; \alpha, \beta)$ becomes $h(x-\alpha, y-\beta)$ in (2.1) and $\mathbf{H}$ takes the form of a block-Toeplitz matrix.

A Toeplitz matrix [2] is a matrix where every element lying on the same diagonal line has the same value. Here is an example of a Toeplitz matrix:

$$\begin{bmatrix} 1 & 2 & 3 & 4 & 5 \\ 2 & 1 & 2 & 3 & 4 \\ 3 & 2 & 1 & 2 & 3 \\ 4 & 3 & 2 & 1 & 2 \\ 5 & 4 & 3 & 2 & 1 \end{bmatrix}$$

A block-Toeplitz matrix is a matrix that can be divided into a number of equal sized blocks. Each block is a Toeplitz matrix, and blocks lying on the same *block diagonal* are identical. Here is an example of a 6 by 6 block-Toeplitz matrix:

$$
\begin{bmatrix}
1 & 2 & 3 & 4 & 5 & 6 \\
2 & 1 & 4 & 3 & 6 & 5 \\
3 & 4 & 1 & 2 & 3 & 4 \\
4 & 3 & 2 & 1 & 4 & 3 \\
5 & 6 & 3 & 4 & 1 & 2 \\
6 & 5 & 4 & 3 & 2 & 1
\end{bmatrix}
=
\begin{bmatrix}
\mathbf{H}_{11} & \mathbf{H}_{22} & \mathbf{H}_{33} \\
\mathbf{H}_{22} & \mathbf{H}_{11} & \mathbf{H}_{22} \\
\mathbf{H}_{33} & \mathbf{H}_{22} & \mathbf{H}_{11}
\end{bmatrix}
$$

where:

$$
\mathbf{H}_{11} = \begin{bmatrix} 1 & 2 \\ 2 & 1 \end{bmatrix}, \mathbf{H}_{22} = \begin{bmatrix} 3 & 4 \\ 4 & 3 \end{bmatrix}, \mathbf{H}_{33} = \begin{bmatrix} 5 & 6 \\ 6 & 5 \end{bmatrix}
$$

Notice that a Toeplitz matrix is also a block-Toeplitz matrix with a block site of 1 by 1, but a block Toeplitz matrix is usually not Toeplitz. The block-Toeplitz structure of $\mathbf{H}$ comes about due to the block structure of $\mathbf{f}$, $\mathbf{g}$ and $\mathbf{n}$ created by the lexicographic ordering. If $h(x, y; \alpha, \beta)$ has a simple form of space variance then $\mathbf{H}$ may have a simple form, resembling a block-Toeplitz matrix.

2.2   Image Restoration

When an image is recorded suffering some type of degradation, such as mentioned above, it may not always be possible to take another, undistorted, image of the interesting phenomena or object. The situation may not recur, like the image of a planet taken by a space probe, or the image of a crime in progress. On the other hand, the imaging system used may introduce inherent distortions to the image which cannot be avoided, for example, a Magnetic Resonance Imaging system. To restore an image degraded by a linear distortion, a restoration cost function can be developed. The cost function is created using knowledge about the degraded image and an estimate of the degradation, and possibly noise, suffered by the original image to produce the degraded image. The free variable in the cost function is an image, that we will denote by $\hat{\mathbf{f}}$, and the cost function is designed such that the $\hat{\mathbf{f}}$ which minimizes the cost function is an estimate of the original image. A common class of cost functions is based on the mean square error (MSE) between the original image and the estimate image. Cost functions based on the MSE often have a quadratic nature.

2.2.1   Degradation Measure

In this work, the degradation measure we consider minimizing starts with the constrained least square error measure [4]:

$$
E = \frac{1}{2}\|\mathbf{g} - \mathbf{H}\hat{\mathbf{f}}\|^2 + \frac{1}{2}\lambda\|\mathbf{D}\hat{\mathbf{f}}\|^2 \tag{2.3}
$$

where $\hat{\mathbf{f}}$ is the restored image estimate, λ is a constant, and $\mathbf{D}$ is a smoothness constraint operator. Since $\mathbf{H}$ is often a low-pass distortion, $\mathbf{D}$ will be chosen to be a high-pass filter. The second term in (2.3) is the regularization term. The more noise that exists in an image, the greater the second term in (2.3) should be, hence minimizing the second term will involve reducing the noise in the image at the expense of restoration sharpness.

Choosing λ becomes an important consideration when restoring an image. Too great a value of λ will oversmooth the restored image, whereas too small a value of λ will not properly suppress noise. At their essence, neural networks minimize cost functions such as that above. It is not unexpected that there exist neural network models to restore degraded imagery.

2.2.2 Neural Network Restoration

Neural network restoration approaches are designed to minimize a quadratic programming problem [46, 105, 106, 107, 108]. The generalized Hopfield Network can be applied to this case [35]. The general form of a quadratic programming problem can be stated as:

Minimize the energy function associated with a neural network given by:

$$E = -\frac{1}{2}\,\hat{\mathbf{f}}^T\mathbf{W}\hat{\mathbf{f}} - \mathbf{b}^T\hat{\mathbf{f}} + c \tag{2.4}$$

Comparing this with (2.3), $\mathbf{W}$, $\mathbf{b}$ and c are functions of $\mathbf{H}$, $\mathbf{D}$, λ and $\mathbf{n}$, and other problem related constraints. In terms of a neural network energy function, the (i, j)th element of $\mathbf{W}$ corresponds to the interconnection strength between neurons (pixels) i and j in the network. Similarly, vector $\mathbf{b}$ corresponds to the bias input to each neuron.

Equating the formula for the energy of a neural network with equation (2.3), the bias inputs and interconnection strengths can be found such that as the neural network minimizes its energy function, the image will be restored.

Expanding (2.3) we get:

$$
\begin{aligned}
E =& \frac{1}{2}\sum_{p=1}^{L}(g_p - \sum_{i=1}^{L} h_{pi}\hat{f}_i)^2 + \frac{1}{2}\lambda\sum_{p=1}^{L}(\sum_{i=1}^{L} d_{pi}\hat{f}_i)^2 \\
=& \frac{1}{2}\sum_{p=1}^{L}(g_p - \sum_{i=1}^{L} h_{pi}\hat{f}_i)(g_p - \sum_{j=1}^{L} h_{pj}\hat{f}_j) + \frac{1}{2}\lambda\sum_{p=1}^{L}(\sum_{i=1}^{L} d_{pi}\hat{f}_i)(\sum_{j=1}^{L} d_{pj}\hat{f}_j) \\
=& \frac{1}{2}\sum_{p=1}^{L}((g_p)^2 - 2g_p\sum_{i=1}^{L} h_{pi}\hat{f}_i + \sum_{i=1}^{L} h_{pi}\hat{f}_i\sum_{j=1}^{L} h_{pj}\hat{f}_j) \\
& + \frac{1}{2}\lambda\sum_{p=1}^{L}(\sum_{i=1}^{L} d_{pi}\hat{f}_i)(\sum_{j=1}^{L} d_{pj}\hat{f}_j) \\
=& \frac{1}{2}\sum_{p=1}^{L}(g_p)^2 - \sum_{p=1}^{L}\sum_{i=1}^{L} g_p h_{pi}\hat{f}_i + \frac{1}{2}\sum_{p=1}^{L}\sum_{i=1}^{L} h_{pi}\hat{f}_i\sum_{j=1}^{L} h_{pj}\hat{f}_j \\
& + \frac{1}{2}\lambda\sum_{p=1}^{L}\sum_{i=1}^{L} d_{pi}\hat{f}_i\sum_{j=1}^{L} d_{pj}\hat{f}_j \\
=& \frac{1}{2}\sum_{p=1}^{L}\sum_{i=1}^{L}\sum_{j=1}^{L} h_{pj}\hat{f}_j h_{pi}\hat{f}_i + \frac{1}{2}\lambda\sum_{p=1}^{L}\sum_{i=1}^{L}\sum_{j=1}^{L} d_{pj}\hat{f}_j d_{pi}\hat{f}_i \\
& - \sum_{p=1}^{L}\sum_{i=1}^{L} g_p h_{pi}\hat{f}_i + \frac{1}{2}\sum_{p=1}^{L}(g_p)^2
\end{aligned}
$$

Hence

$$
E = \frac{1}{2}\sum_{i=1}^{L}\sum_{j=1}^{L}\left(\sum_{p=1}^{L} h_{pj}h_{pi} + \lambda\sum_{p=1}^{L} d_{pj}d_{pi}\right)\hat{f}_i\hat{f}_j - \sum_{i=1}^{L}\sum_{p=1}^{L} g_p h_{pi}\hat{f}_i + \frac{1}{2}\sum_{p=1}^{L}(g_p)^2
\tag{2.5}
$$

Expanding (2.4) we get:

$$
E = -\frac{1}{2}\sum_{i=1}^{L}\sum_{j=1}^{L} w_{ij}\hat{f}_i\hat{f}_j - \sum_{i=1}^{L} b_i\hat{f}_i + c
\tag{2.6}
$$

By equating the terms in equations (2.5) and (2.6) we find that the neural network model can be matched to the constrained least square error cost function by ignoring the constant, c, and setting:

$$
w_{ij} = -\sum_{p=1}^{L} h_{pi}h_{pj} - \lambda\sum_{p=1}^{L} d_{pi}d_{pj}
\tag{2.7}
$$

and

$$b_i = \sum_{p=1}^{L} g_p h_{pi} \tag{2.8}$$

where w_{ij} is the interconnection strength between pixels i and j, and b_i is the bias input to neuron (pixel) i. In addition, h_{ij} is the (i, j)th element of matrix $\mathbf{H}$ from equation (2.3) and d_{ij} is the (i, j)th element of matrix $\mathbf{D}$.

Now let's look at some neural networks in the literature to solve this problem.

2.3 Neural Network Restoration Algorithms in the Literature

In the network described by Zhou *et al.* For an image with $S + 1$ gray levels, each pixel is represented by $S + 1$ neurons [46]. Each neuron can have a value of 0 or 1. The value of the ith pixel is then given by:

$$\hat{f}_i = \sum_{k=0}^{S} v_{i,k} \tag{2.9}$$

where $v_{i,k}$ is the state of the kth neuron of the ith pixel. Each neuron is visited sequentially and has its input calculated according to:

$$u_i = b_i + \sum_{j=1}^{L} w_{ij} \hat{f}_j \tag{2.10}$$

where u_i is the input to neuron i, and $\hat{f}_i$ is the state of the jth neuron. Based on u_i, the neuron's state is updated according to the following rule:

$$\Delta \hat{f}_i = G(u_i)$$

where

$$G(u) = \begin{cases} 1, & u > 0 \\ 0, & u = 0 \\ -1, & u < 0 \end{cases} \tag{2.11}$$

The change in energy resulting from a change in neuron state of $\Delta \hat{f}_i$ is given by:

$$\Delta E = -\frac{1}{2} w_{ii} (\Delta \hat{f}_i)^2 - u_i \Delta \hat{f}_i \tag{2.12}$$

If $\Delta E < 0$, then the neuron's state is updated. This algorithm may be summarized as:

Algorithm 2.1:

repeat
{
    For $i = 1, \ldots, L$ do
    {
        For $k = 0, \ldots, S$ do
        {
$$u_i = b_i + \sum_{j=1}^{L} w_{ij} \hat{f}_j$$
$$\Delta \hat{f}_i = G(u_i)$$
$$\text{where } G(u) = \begin{cases} 1, & u > 0 \\ 0, & u = 0 \\ -1, & u < 0 \end{cases}$$
$$\Delta E = -\tfrac{1}{2} w_{ii} (\Delta \hat{f}_i)^2 - u_i \Delta \hat{f}_i$$
If $\Delta E < 0$, then $v_{i,k} = v_{i,k} + \Delta \hat{f}_i$
$$\hat{f}_i = \sum_{k=0}^{S} v_{i,k}$$
        }
    }
    $t = t + 1$
}
until $\hat{f}_i(t) = \hat{f}_i(t-1) \forall i = 1, \ldots, L)$

In the paper by Paik and Katsaggelos, Algorithm 2.1 was enhanced to re-move the step where the energy reduction is checked following the calculation of $\Delta \hat{f}_i$ [105]. Paik and Katsaggelos presented an algorithm which made use of a more complicated neuron. In their model, each pixel was represented by a single neuron which takes discrete values between 0 and S, and is capable of updating its value by ± 1, or keeping the same value during a single step. A new method for calculating $\Delta \hat{f}_i$ was also presented:

$$\Delta \hat{f}_i = \acute{G}_i(u_i)$$

where

$$\acute{G}_i = \begin{cases} -1, & u < -\theta_i \\ 0, & -\theta_i \leq u \leq \theta_i \\ 1, & u > \theta_i \end{cases} \tag{2.13}$$

where $\theta_i = -\tfrac{1}{2} w_{ii} > 0$.

This algorithm may be presented as:

Algorithm 2.2:

repeat
{

   For $i = 1, \ldots, L$ do
   {

$$u_i = b_i + \sum_{j=1}^{L} w_{ij} \hat{f}_j$$
$$\Delta \hat{f}_i = \acute{G}_i(u_i)$$
$$\text{where } \acute{G}_i(u) = \begin{cases} -1, & u < -\theta_i \\ 0, & -\theta_i \leq u \leq \theta_i \\ 1, & u > \theta_i \end{cases}$$
$$\text{where } \theta_i = -\tfrac{1}{2} w_{ii} > 0$$
$$\hat{f}_i(t+1) = K(\hat{f}_i(t) + \Delta \hat{f}_i)$$
$$\text{where } K(u) = \begin{cases} 0, & u < 0 \\ u, & 0 \leq u \leq S \\ S, & u \geq S \end{cases}$$

   }
   $t = t + 1$

}
until $\hat{f}_i(t) = \hat{f}_i(t-1) \forall i = 1, \ldots, L$)

Algorithm 2.2 makes no specific check that energy has decreased during each iteration and so in [105] they proved that Algorithm 2.2 would result in a decrease of the energy function at each iteration. Note that in Algorithm 2.2, each pixel only changes its value by ± 1 during an iteration. In Algorithm 2.1, the pixel's value would change by any amount between 0 and S during an iteration since each pixel was represented by $S + 1$ neurons. Although Algorithm 2.2 is much more efficient in terms of the number of neurons used, it may take many more iterations than Algorithm 2.1 to converge to a solution (although the time taken may still be faster than Algorithm 2.1). If we consider that the value of each pixel represents a dimension of the L dimensional energy function to be minimized, then we can see that Algorithms 2.1 and 2.2 have slightly different approaches to finding a local minimum. In Algorithm 2.1, the energy function is minimized along each dimension in turn. The image can be considered to represent a single point in the solution space. In Algorithm 2.1, this point moves to the function minimum along each of the L axes of the problem until it eventually reaches a local minimum of the energy function. In Algorithm 2.2, for each pixel, the point takes a unit step in a direction that reduces the network energy along that dimension. If the weight matrix is negative definite ($-\mathbf{W}$ is positive definite), however, regardless of how these algorithms work, the end results must be similar (if each algorithm ends at a minimum). The reason for this is that when the weight matrix is negative definite, there is only the global minimum. That is, the function has only one minimum. In this case the matrix $\mathbf{W}$ is invertible and taking (2.4) we see that:

$$\frac{\delta E}{\delta \hat{\mathbf{f}}} = -\mathbf{W}\hat{\mathbf{f}} - \mathbf{b} \tag{2.14}$$

Hence the solution is given by:

$$\hat{\mathbf{f}}^{\star} = -\mathbf{W}^{-1}\mathbf{b} \tag{2.15}$$

(assuming that $\mathbf{W}^{-1}$ exists).

The $\hat{\mathbf{f}}^{\star}$ is the only minimum and the only stationary point of this cost function, so we can state that if $\mathbf{W}$ is negative definite and Algorithm 2.1 and Algorithm 2.2 both terminate at a local minimum, the resultant image must be close to $\hat{\mathbf{f}}^{\star}$ for both algorithms. Algorithm 2.1 approaches the minimum in a zigzag fashion, whereas Algorithm 2.2 approaches the minimum with a smooth curve. If $\mathbf{W}$ is not negative definite, then local minimum may exist and Algorithms 2.1 and 2.2 may not produce the same results. If Algorithm 2.2 is altered so that instead of changing each neuron's value by ± 1 before going to the next neuron, the current neuron is iterated until the input to that neuron is zero, then Algorithms 2.1 and 2.2 will produce identical results. Each algorithm will terminate in the same local minimum.

2.4   An Improved Algorithm

Although Algorithm 2.2 is an improvement on Algorithm 2.1, it is not optimal. From iteration to iteration, neurons often oscillate about their final value, and during the initial iterations of Algorithm 2.1 a neuron may require 100 or more state changes in order to minimize its energy contribution. A faster method to minimize the energy contribution of each neuron being considered is suggested by examination of the mathematics involved. For an image where each pixel is able to take on any discrete integer intensity between 0 and S, we assign each pixel in the image to a single neuron able to take any discrete value between 0 and S. Since the formula for the energy reduction resulting from a change in neuron state $\Delta \hat{f}_i$ is a simple quadratic, it is possible to solve for the $\Delta \hat{f}_i$ which produces the maximum energy reduction. Theorem 2.1 states that this approach will result in the same energy minimum as Algorithm 2.1 and hence the same final state of each neuron after it is updated.

Theorem 2.1: *For each neuron i in the network during each iteration, there exists a state change $\Delta \hat{f}_i^{*}$ such that the energy contribution of neuron i is minimized.*

Proof:

Let u_i be the input to neuron i which is calculated by:

$$u_i = b_i + \sum_{j=1}^{L} w_{ij} \hat{f}_j$$

Let ΔE be the resulting energy change due to $\Delta \hat{f}_i$.

$$\Delta E = -\frac{1}{2} w_{ii} (\Delta \hat{f}_i)^2 - u_i \Delta \hat{f}_i \tag{2.16}$$

Differentiating ΔE with respect to $\Delta \hat{f}_i$ gives us:

$$\frac{\delta \Delta E}{\delta \hat{f}_i} = -w_{ii} \Delta \hat{f}_i - u_i$$

The value of $\Delta \hat{f}_i$ which minimizes (2.16) is given by:

$$0 = -w_{ii} \Delta \hat{f}_i^* - u_i$$

Therefore,

$$\Delta \hat{f}_i^* = \frac{-u_i}{w_{ii}} \tag{2.17}$$

QED.

Based on Theorem 2.1, an improved algorithm is presented below.

Algorithm 2.3.

repeat
{
    For $i = 1, \ldots, L$ do
    {

$$u_i = b_i + \sum_{j=1}^{L} w_{ij} \hat{f}_j$$
$$\Delta \hat{f}_i = G(u_i)$$

$$\text{where } G(u) = \begin{cases} -1, & u < 0 \\ 0, & u = 0 \\ 1, & u > 0 \end{cases}$$

$$\Delta E_{ss} = -\frac{1}{2} w_{ii} (\Delta \hat{f}_i)^2 - u_i \Delta \hat{f}_i \tag{2.18}$$

If $\Delta E_{ss} < 0$ then $\Delta \hat{f}_i^* = \dfrac{-u_i}{w_{ii}}$
$$\hat{f}_i(t+1) = K(\hat{f}_i(t) + \Delta \hat{f}_i^*)$$
$$\text{where } K(u) = \begin{cases} 0, & u < 0 \\ u, & 0 \leq u \leq S \\ S, & u \geq S \end{cases}$$

}

$$t = t + 1$$
}
until $\hat{f}_i(t) = \hat{f}_i(t-1) \forall i = 1, \ldots, L)$

Each neuron is visited sequentially and has its input calculated. Using the input value, the state change needed to minimize the neuron's energy contribution to the network is calculated. Note that since $\Delta \hat{f}_i \in \{-1, 0, 1\}$ and $\Delta \hat{f}_i$ and $\Delta \hat{f}_i^*$ must be the same sign as u_i, step (2.18) is equivalent to checking that at least a unit step can be taken which will reduce the energy of the network. If $\Delta E_{ss} < 0$, then

$$-\tfrac{1}{2}w_{ii} - u_i \Delta \hat{f}_i < 0$$
$$-\tfrac{1}{2}w_{ii} - |u_i| < 0$$
$$-w_{ii} < 2|u_i|$$

Substituting this result into the formula for $\Delta \hat{f}_i^*$ we get:

$$\Delta \hat{f}_i^* = \frac{-u_i}{w_{ii}} > \frac{u_i}{2|u_i|} = \frac{1}{2}\Delta \hat{f}_i$$

Since $\Delta \hat{f}_i^*$ and $\Delta \hat{f}_i$ have the same sign and $\Delta \hat{f}_i = \pm 1$ we obtain:

$$|\Delta \hat{f}_i^*| > \frac{1}{2} \tag{2.19}$$

In this way, $\Delta \hat{f}_i^*$ will always be large enough to alter the neuron's discrete value.

Algorithm 2.3 makes use of concepts from both Algorithm 2.1 and Algorithm 2.2. Like Algorithm 2.1 the energy function is minimized in solution space along each dimension in turn until a local minimum is reached. In addition, the efficient use of space by Algorithm 2.2 is utilized. Note that the above algorithm is much faster than either Algorithm 2.1 or 2.2 due to the fact that this algorithm minimizes the current neuron's energy contribution in one step rather than through numerous iterations as did Algorithms 2.1 and 2.2.

2.5 Analysis

In the paper by Paik and Katsaggelos, it was shown that Algorithm 2.2 would converge to a fixed point after a finite number of iterations and that the fixed point would be a local minimum of E in (2.3) in the case of a sequential algorithm [105]. Here we will show that Algorithm 2.3 will also converge to a fixed point which is a local minimum of E in (2.3).

Algorithm 2.2 makes no specific check that energy has decreased during each iteration and so in [105] they proved that Algorithm 2.2 would result in a decrease of the energy function at each iteration. Algorithm 2.3, however, changes the current neuron's state if and only if an energy reduction will occur and $|\Delta \hat{f}_i| = 1$. For this reason Algorithm 2.3 can only reduce the energy function and never increase it. From this we can observe that each iteration of Algorithm 2.3 brings the network closer to a local minimum of the function. The next question is *Does Algorithm 2.3 ever reach a local minimum and terminate?* Note that the gradient of the function is given by:

$$\frac{\delta E}{\delta \hat{\mathbf{f}}} - \mathbf{W}\hat{\mathbf{f}} - \mathbf{b} = -\mathbf{u} \tag{2.20}$$

where $\mathbf{u}$ is a vector whose ith element contains the current input to neuron i. Note that during any iteration, $\mathbf{u}$ will always point in a direction that reduces the energy function. If $\hat{\mathbf{f}} \neq \hat{\mathbf{f}}^\star$ then for at least one neuron a change in state must be possible which would reduce the energy function. For this neuron, $u_i \neq 0$. The algorithm will then compute the change in state for this neuron to move closer to the solution. If $|\Delta \hat{f}_i^*| > \frac{1}{2}$ the neuron's state will be changed. In this case we assume that no boundary conditions have been activated to stop neuron i from changing value. Due to the discrete nature of the neuron states we see that the step size taken by the network is never less than 1.

To restate the facts obtained so far:

- During each iteration Algorithm 2.3 will reduce the energy of the network.

- A reduction in the energy of the network implies that the network has moved closer to a local minimum of the energy function.

- There is a lower bound to the step size taken by the network and a finite range of neuron states. Since the network is restricted to changing state only when an energy reduction is possible, the network cannot iterate forever.

From these observations we can conclude that the network reaches a local minimum in a finite number of iterations, and that the solution given by Algorithm 2.3 will be close to the solution given by Algorithm 2.1 for the same problem. The reason Algorithms 2.1 and 2.3 must approach the same local minimum is the fact that they operate on the pixel in an identical manner. In Algorithm 2.1 each of the $S + 1$ neurons associated with pixel i is adjusted to reduce its contribution to the energy function. The sum of the contributions of the $S + 1$ neurons associated with pixel i in Algorithm 2.1 equals the final grayscale value of that neuron. Hence during any iteration of Algorithm 2.1 the current pixel can change to any allowable value. There are $S + 1$ possible output values of pixel i and only one of these values results when the algorithm minimizes the contribution of that pixel. Hence whether the pixel is represented by $S + 1$ neurons or just a single neuron, the output grayscale value that occurs when the energy contribution of that pixel is minimized during the current iteration remains the same. Algorithms 2.1 and 2.3 both minimize the current pixel's energy contribution; hence they must both produce the same results. In practice the authors have found that all three algorithms generally produce identical results, which suggests that for reasonable values of the parameter λ, only a single global minimum is present.

Note that in the discussion so far, we have not made any assumptions regarding the nature of the weighting matrix, $\mathbf{W}$, or the bias vector, $\mathbf{b}$. $\mathbf{W}$ and $\mathbf{b}$ determine where the solution is in the solution space, but as long as they are constant during the restoration procedure the algorithm will still terminate after a finite number of iterations. This is an important result, and implies that even if the degradation suffered by the image is space variant or if we assign a different value of λ to each pixel in the image, the algorithm will still converge to a result. Even if $\mathbf{W}$ and $\mathbf{b}$ are such that the solution lies outside of the bounds on the values of the neurons, we would still expect that there exists a point or points which minimize E within the bounds. In practice we would not expect the solution to lie entirely out of the range of neuron values. If we assume that Algorithm 2.3 has terminated at a position where no boundary conditions have been activated, then the condition:

$$\left| \Delta \hat{f}_i^* \right| = \left| \frac{-u_i}{w_{ii}} \right| < \frac{1}{2}, \forall i \in \{0, 1, \ldots, L\}$$

must have been met. This implies that:

$$|u_i| < \frac{1}{2} w_{ii}, \forall i \in \{0, 1, \ldots, L\} \qquad (2.21)$$

In [105], Paik and Katsaggelos noticed this feature as well, since the same termination conditions apply to Algorithm 2.2. The self-connection weight, w_{ii}, controls how close to the ideal solution the algorithm will approach before terminating. Since increasing the value of λ increases the value of w_{ii}, we would expect also that the algorithm would terminate more quickly and yet be less accurate for larger values of λ. This is found to occur in practice. When λ is increased, the number of iterations before termination drops rapidly.

2.6   Implementation Considerations

Despite the increase in efficiency and speed of Algorithm 2.3 when compared to Algorithms 2.1 and 2.2, there are still a number of ways that the algorithm can be made more efficient. The ith row of the weighting matrix describes the interconnection strengths between neuron i and every other neuron in the network from the *viewpoint* of neuron i. The weighting matrix is NM by NM, which is clearly a prohibitively large amount of data that requires storage. However, the mathematical discussion in the previous sections implies a shortcut.

By examining (2.7) we observe that in the case of $P \ll \min(M, N)$, it can be seen that when calculating the input to each neuron, only pixels within a certain rectangular neighborhood of the current pixel contribute non-zero components to the neuron input. In addition, it can be seen that the non-zero interconnection strengths between any given pixel and a neighboring pixel depend only on the position of the pixels relative to each other in the case of spatially invariant distortion. Using the above observations, the input to any neuron (pixel) in the image can be calculated by applying a mask to the image centered on the pixel being examined. For a P by P distortion, each weighting mask contains only $(2P - 1)^2$ terms. A 5 by 5 degrading PSF acting on a 250 by 250 image requires a weight matrix containing 3.9×10^9 elements, yet a weighting mask of only 81 elements. In addition, by considering the finite regions of support of the degrading and filtering impulse functions represented by $\mathbf{H}$ and $\mathbf{D}$, the weighting masks and bias inputs to each neuron may be calculated without storing matrices $\mathbf{H}$ and $\mathbf{D}$ at all. They may be calculated using only the impulse responses of the above matrices.

2.7 A Numerical Study of the Algorithms

To compare the three algorithms, let us look at two examples. In the first example, the efficiency of Algorithms 2.1, 2.2 and 2.3 will be compared to one another. In the second example, a practical example of the use of this method will be given.

2.7.1 Setup

In both these examples, the images were blurred using a Gaussian PSF with the impulse response:

$$h(x, y) = \frac{1}{2\pi\sigma_x\sigma_y} \exp\left[-\left(\frac{x^2}{2\sigma_x^2} + \frac{y^2}{2\sigma_y^2}\right)\right] \tag{2.22}$$

where σ_x and σ_y are the standard deviations of the PSF in the x and y directions, respectively.

2.7.2 Efficiency

The time taken to restore an image was compared among Algorithms 2.1, 2.2 and 2.3. A degraded image was created by blurring a 256 by 256 image with a Gaussian blur of size 5 by 5 and standard deviation 2.0. Noise of variance 4.22 was added to the blurred image. Each algorithm was run until at least 85% of the pixels in the image no longer changed value or many iterations had passed with the same number of pixels changing value during each iteration. Algorithm 2.1 was stopped after the sixth iteration when no further improvement was possible, and took 6067 seconds to run on a SUN Ultra SPARC 1. Algorithm 2.2 was stopped after the 30th iteration with 89% of pixels having converged to their stable states and took 126 seconds to run. Algorithm 2.3 was stopped after the 18th iteration with 90% of pixels stable and took 78 seconds to run. Algorithm 2.3 is much faster than Algorithms 2.1 and 2.2, despite the fact that algorithms 2.1 and 2.3 approach the same local minimum and hence give the same results. The computation time of Algorithm 2.3 can be expected to increase linearly with the number of pixels in the image, as can the computation times of Algorithms 2.1 and 2.2. The single step neuron energy minimization technique of Algorithm 2.3 provides its superior speed and this trend holds for any size image. Various types of distortions and noise would not be expected to change the speed relationship between Algorithms 2.1, 2.2 and 2.3 for any given image. This is because each algorithm was shown to converge to similar points in solution space and instead use different methods to reach this point.

2.7.3   An Application Example

Algorithm 2.3 was applied to the problem of restoring an image with an unknown level of motion blur. Restoring an image distorted by motion blur is a very common problem.

Figure 2.1a shows an aircraft wing in a state of assembly. The camera had shifted with an unknown degree of movement during some of the exposures thereby blurring the images.

Although the image appears to be blurred by motion from left to right, it was desired to confirm this by restoring the image using the network parameters derived from motion in horizontal, vertical and both diagonal directions. The

Figure 2.1: Degraded and restored wing images

neural network algorithm is perfectly suited to rapid investigations because different regions of the image can be set to be restored using the parameters of different point spread functions. Figure 2.1b shows Figure 2.1a restored using both diagonal point spread functions (top left and bottom right of the image), a vertical motion blur point spread function (bottom left of the image) and a horizontal motion blur point spread function (top right of the image). Using the image partitioning ability of this algorithm [109], four exploratory restorations were performed simultaneously. As a result of this, Figure 2.1b demonstrates that the best matching point spread function was that associated with horizontal camera movement, since the top right section of the image is clearly the most well restored.

The second part of the investigation was to determine whether the horizontal motion was from left to right, right to left or due to camera shake rather than motion blur (in which case the motion will be in both directions). Figure 2.1c shows Figure 2.1a restored using the parameters of four different point spread functions. The top left of the image was restored using the parameters of a uniform point spread function (as a control). The bottom left of the image was restored with a point spread function associated with camera shake. The top right and bottom right regions were restored using the parameters of horizontal motion blur to the right and left, respectively. Figure 2.1c shows that the image was degraded with horizontal motion blur to the right as this set of network parameters produces the sharpest restoration. Figure 2.1d shows the final restored image using the optimal point spread function determined previously.

2.8 Summary

In this chapter, we have examined the basic neural network optimization algorithm. We first looked at the problem of restoring an image acted upon by a degrading PSF in the presence of noise. The problem of reversing the degrading process and approximating the original image can be formulated as a minimization process on a constrained least square error measure function. This error measure can be minimized by relating the terms of the error measure with the formula for the energy of a Hopfield-style neural network. By selecting suitable interconnection strengths (weights) and bias inputs for the neurons, we can equate the neural network energy function with the constrained least square error cost function. The problem then reduces to selecting the optimal algorithm to minimize the network energy function. We examined two such algorithms presented previously in the literature and considered the advantages and disadvantages of both. We then presented a new algorithm which brings together features from both the previous algorithms. The new algorithm was shown to converge to a solution in a finite number of steps with much faster speed than either of the previous algorithms. We looked at some interesting properties of the algorithm including the fact that the convergence of the algorithm was irrespective of whether the degradation was spatially invariant or spatially variant. We examined implementation considerations involved with using the algorithm in practice and number of examples were given. The examples showed that the proposed algorithm is much faster than previous algorithms and a practical example of the use of the algorithm was given.

Chapter 3

Spatially Adaptive Image Restoration

3.1 Introduction

In the previous chapter, it was seen that as long as the weights of the network remain constant from iteration to iteration, Algorithm 2.3 will converge. This result holds regardless of whether the weights of the network vary from pixel to pixel across the image to be restored. We will refer to this type of variation of the weights from region to region in the image as *spatial variation*.

The analysis in the previous chapter focused solely on the case when the weights are not spatially variant. But there are situations when we may want to solve problems that involve spatially variant weights. The first case occurs when the type and/or strength of the degrading function varies from region to region in the image. For example, imagine taking a picture of a distant mountain when a nearby tree is in the picture. If the image of the mountain is in focus, then the image of the tree will most likely be out of focus. The pixels belonging to the image of the tree have suffered a more severe distortion than the pixels belonging to the image of the mountain. We will term this type of effect as *spatially variant distortion*. In spatially variant distortion, the columns of the degradation matrix, **H**, that correspond to pixels in the image of the tree will be different from the rest of the degradation matrix. In turn, this will cause the weights assigned to the neurons associated with these pixels to also be different from the rest of the neurons in the network. Spatially variant distortion may also be caused by moving objects in the picture or local problems with the imaging system such as impurities in the camera lens. In fact, spatially variant distortion is the most common type of degradation occurring in optical imagery. The problem of varying the restoration weights to handle spatially variant distortion will be examined in this chapter and examples given.

The second case of when we may want to vary the weights of the network spatially is more fundamental than the case of spatially variant distortion. The restoration model we developed in Chapter 2 has a single parameter to control high frequency components in the image and balance the desire to restore the image against the desire to reduce noise in the image. But this approach makes the assumption that in all regions of the image, the balance is approximately the same. That is, all parts of the image have the same statistical properties of the signal and noise. Real images almost always violate this assumption. Most images of interesting objects or phenomena contain many different regions with different statistical properties. For example, real images contain edges, smooth areas, textures of different types and fine and coarse details. Each of these areas has its own optimal balance between restoration and noise suppression, which may be different from other areas. It seems that the best way to handle this problem would be to vary the restoration parameter from pixel to pixel or region to region.

In *adaptive constraint restoration*, the weights are varied to implement different values of regularization parameter for different regions of the image. By doing this, we can adjust restoration parameters to suppress noise more greatly in regions where it is most noticeable, and less so in regions where image sharpness is the dominant consideration. There are two ways to look at this problem.

One way to look at this problem is to think of the Constrained Least Square Error formula (2.3) as containing a restoration term and a penalty term. The natural approach is then to assign each neuron in the network its own regularization parameter rather than letting all neurons have the same value. Then the problem becomes how to select the best value of regularization parameter for each pixel. This may be done using energy considerations with a gradient descent algorithm, or the statistical properties of the local region of the neuron. In this chapter we will analyze the penalty function concept and show that the method for selecting the regularization parameter based on gradient descent can find only sub-optimal solutions. We also mathematically verify a fact, which has been observed in practice, that this method would generally use a small value of regularization parameter for textured regions, and a large value for smooth regions. Using this observation, we then introduce a regional processing approach based on local statistics to select the optimal regularization parameter for each pixel. This approach also has some relationship to biological vision systems in that it emphasizes on edges. The algorithms described in the previous chapter are expanded to vary the neural weights to take account of a spatially variant PSF and to combine the adaptive constraint concept developed. We then look at how adaptive constraint restoration can be used to compensate for incomplete knowledge of the degrading PSF or incomplete knowledge of the spatial variance of the degradation.

The second way to look at this problem is to consider the Constrained Least Square Error formula (2.3) as being a Lagrangian formulation of the restoration problem. In this chapter we will examine this concept and show that the natural extension to adaptive constraint restoration is to divide the image into regions, with a value of regularization parameter assigned to each region rather than each pixel.

This chapter is divided into a number of sections. Section 3.2 examines the problem of handling spatially variant distortions, while Section 3.3 introduces the adaptive constraint algorithms derived from the penalty function concept. Section 3.4 describes how adaptive constraint restoration can be merged with methods to handle spatially variant distortions while Section 3.5 considers the problem of semi-blind restoration. This is restoration when only incomplete knowledge of the degrading function is available. Section 3.6 details some implementation considerations, while Section 3.7 presents examples to showcase the concepts presented in the previous sections and Section 3.8 briefly describes the concept behind the Lagrange approach to adaptive constraint restoration. Section 3.9 summarizes this chapter.

3.2 Dealing with Spatially Variant Distortion

In the analysis in Chapter 2, no conditions were placed upon the form of the matrix $\mathbf{H}$. In fact the neural network cost function only assumes that the weighting matrix, $\mathbf{W}$, is symmetric. In the case of a space invariant degradation, the matrix $\mathbf{H}$ will be symmetric. However, in the real world this is not often the case. As discussed in Chapter 2, the general form of an image degradation will be space variant. However, many of these distortions will still be linear and so the model given by equation (2.2) still holds. In the case of a linear space variant distortion, $\mathbf{H}$ will not be symmetric, but the weighting matrix, $\mathbf{W}$, will still be symmetric. This can be clearly shown by examining equation describing the neural weights:

$$w_{ij} = -\sum_{p=1}^{L} h_{pi}h_{pj} - \lambda \sum_{p=1}^{L} d_{pi}d_{pj} \tag{3.1}$$

By converting (3.1) into matrix notation we see that the weighting matrix is given by:

$$\mathbf{W} = -\mathbf{H}^T\mathbf{H} - \lambda\mathbf{D}^T\mathbf{D} \tag{3.2}$$

Assume that $\mathbf{H}$ is not symmetric, such that $\mathbf{H} \neq \mathbf{H}^T$; then:

$$\begin{aligned}
\mathbf{W}^T &= -\left(\mathbf{H}^T\mathbf{H} + \lambda\mathbf{D}^T\mathbf{D}\right)^T \\
&= -\left(\mathbf{H}^T\mathbf{H}\right)^T - \lambda\left(\mathbf{D}^T\mathbf{D}\right)^T \\
&= -\mathbf{H}^T(\mathbf{H}^T)^T - \lambda\mathbf{D}^T(\mathbf{D}^T)^T \\
&= -\mathbf{H}^T\mathbf{H} - \lambda\mathbf{D}^T\mathbf{D} = \mathbf{W}
\end{aligned} \tag{3.3}$$

Hence the weighting matrix, $\mathbf{W}$, is symmetric. This is a powerful feature of the neural network approach. If the image suffers a known form of space variance, then the weights of the neurons in the network can be adjusted to restore the image exactly with very little additional computational overhead.

If the degrading PSFs follow some regular or cyclic pattern then further optimizations can be made to the algorithm. Instead of developing the full non-symmetric version of $\mathbf{H}$ for the entire image, certain regular patterns of space variance allow us to compute a set number of simpler alternative versions of the $\mathbf{H}$ matrix which describe each different type of degrading PSF the image has suffered. From this we can compute a number of alternative versions of the $\mathbf{W}$ matrix. Hence we have a number of different sets of weights to select from when restoring the image. In the worst case, completely random spatial variance, there are NM unique sets of weights to restore the image. However, if patterns of spatial variance can be discovered, the number of unique sets of weights can be significantly reduced. Let us analyze the cyclic spatially variant case in detail. Consider a cyclic spatially variant distortion obtained by the use of V PSFs $h_0(x, y), \ldots, h_{V-1}(x, y)$. The pixels in any one row of the image are acted upon by the same PSF; however, the PSF applied to each row is varied cyclically through the sequence:

$$S_H = \{h_0(x, y), h_1(x, y), ..., h_{V-1}(x, y), h_{V-2}(x, y), ..., h_1(x, y)\} \qquad (3.4)$$

The sequence S_H has a period of $2V - 2$, and hence $2V - 2$ unique sets of weights are required to restore the image. This type of distortion is similar to the degradations involved with side-scan sonar images. The method employed in [108, 110] to handle spatially variant distortion was to precompute a number of sets of weights to handle the various PSFs degrading the image. Since the spatial variation of the PSFs was known, the correct set of weights could be chosen to restore the image accurately. Let us look now at an example of restoring an image degraded by a cyclic distortion.

An image was created using a cyclic variation of 7 by 7 Gaussian PSFs. Each PSF had the impulse response:

$$h(x, y) = \frac{1}{2\pi\sigma_x\sigma_y} \exp\left[-\left(\frac{x^2}{2\sigma_x^2} + \frac{y^2}{2\sigma_y^2}\right)\right] \qquad (3.5)$$

where σ_x and σ_y are the standard deviations of the PSF in the x and y directions, respectively.

Using the analysis in this section, V was set to be 4. Table 3.1 details the degrading PSFs used to blur the image as per equation (3.4):

The original image is shown in Figure 3.1a and the degraded image is shown in Figure 3.1b. This image was restored using two techniques. The first technique was a spatially invariant approach. The spatially variant distortion was approximated as a spatially invariant distortion by using a 7 by 7 Gaussian PSF

Table 3.1: Degrading PSFs

Standard Deviation	1.5	2.0	3.0	4.0
PSF	$h_0(x,y)$	$h_1(x,y)$	$h_2(x,y)$	$h_3(x,y)$

Figure 3.1: Original, degraded and restored images suffering spatially variant distortion

of standard deviation 2.55. That is, all pixels were acted upon by one weighting mask, whose components were calculated by approximating the space variant distortion as a space invariant distortion. Since the space variant distortion was very severe, the approximation caused instability in the restored image. The time to restore this image using the spatially invariant approximation was 230 seconds on a SUN Ultra SPARC 1. This image is shown in Figure 3.1c. The second technique was the proposed spatially variant approach with the six correctly calculated weighting masks. This restoration not only is better than Figure 3.1c, but recovers almost all the fine details lost in distortion. The time to restore this image using the spatially variant method was 314 seconds on a SUN Ultra SPARC 1. This image is shown in Figure 3.1d. Note that using Algorithm 2.3, the image could be restored using the correct spatially variant weights with only an extra 36% time penalty when compared to the spatially invariant approximation.

3.3   Adaptive Constraint Extension of the Penalty Function Model

Now we will consider the adaptation of the constraint factor to take account of the non-stationary nature of the image. We will analyze the penalty function model of the restoration problem and develop algorithms based on that model.

3.3.1   Motivation

The basic non-adaptive restoration cost function is given by:

$$E = \frac{1}{2}\|\mathbf{g} - \mathbf{H}\hat{\mathbf{f}}\|^2 + \frac{1}{2}\lambda\|\mathbf{D}\hat{\mathbf{f}}\|^2 \tag{3.6}$$

In the penalty function model of this problem, the first term in (3.6) attempts to control the fidelity of the image restoration, while the second term penalizes image estimates that contain too much noise. In (3.6), a single value of the constraint parameter, λ, is used for every pixel in the image. This formula assumes that the statistics of the image do not change from region to region. Such a process is called *stationary*.

Images are, by nature, ensembles of non-stationary processes. Hence they do not fit the model given by (3.6). For this reason, solutions based on this model can only produce a sub-optimal restoration. An optimal restoration may be achieved by treating statistically dissimilar regions of an image with different restoration strategies or parameters. One method to achieve this is by using an adaptive regularization scheme. When implementing an adaptive regularization parameter scheme, an important consideration is on what basis should the regularization value be determined. In this section, we study the space invariant case. The result will be generalized to the space variant case in the next section.

To substantiate this issue, we first generalize the quadratic model in (3.6) to

$$E = \frac{1}{2}\|\mathbf{g} - \mathbf{H}\hat{\mathbf{f}}\|^2 + \frac{1}{2}\|\sqrt{\mathbf{\Lambda}}\mathbf{D}\hat{\mathbf{f}}\|^2 \tag{3.7}$$

where

$$\sqrt{\mathbf{\Lambda}} = \begin{bmatrix} \sqrt{\lambda_1} & 0 & 0 \\ 0 & \ldots & 0 \\ 0 & 0 & \sqrt{\lambda_{NM}} \end{bmatrix} \tag{3.8}$$

is a diagonal matrix to reflect the adaptive processing nature. When $\lambda_1 = \lambda_2 = \ldots = \lambda_{NM}$, equation (3.7) is reduced to the conventional constrained least squares formula given in (3.6).

By relating (3.7) to the formula for the energy of a neural network, the neural weight between neurons i and j in the network is given by:

$$w_{ij} = -\sum_{p=1}^{L} h_{pi}h_{pj} - \sum_{p=1}^{L} \lambda_p d_{pi}d_{pj} \tag{3.9}$$

Equation (3.9) is overly complicated and can be approximated as:

$$w_{ij} = -\sum_{p=1}^{L} h_{pi}h_{pj} - \lambda_i \sum_{p=1}^{L} d_{pi}d_{pj} \qquad (3.10)$$

Note that the matrix $\mathbf{W}$, whose (i, j)th element is given by (3.10), is not symmetric. As long as a check is made on whether the energy will be decreased before updating a pixel value, the lack of symmetry will not cause the algorithm to fail to converge. This is because $\mathbf{W}$ may be replaced by a symmetric equivalent version, $\mathbf{W}^{\star}$, without changing the energy function.

Define $\mathbf{W}^{\star} = \frac{(\mathbf{W}+\mathbf{W}^T)}{2}$, where $\mathbf{W}$ is the non-symmetric weighting matrix whose (i, j)th element is given by (3.10). Note $\mathbf{W}^{\star}$ is symmetric since:

$$(\mathbf{W}^{\star})^T = \frac{(\mathbf{W}+\mathbf{W}^T)^T}{2} = \frac{(\mathbf{W}^T+\mathbf{W})}{2} = \mathbf{W}^{\star}$$

Taking the formula for the energy of the neural network, and noting that $a^T = a$ when a is a scalar, then we obtain:

$$E = -\frac{1}{2}\,\hat{\mathbf{f}}^T\mathbf{W}^{\star}\hat{\mathbf{f}} - \mathbf{b}^T\hat{\mathbf{f}} + c$$

$$= -\frac{1}{2}\,\hat{\mathbf{f}}^T\frac{(\mathbf{W}+\mathbf{W}^T)}{2}\hat{\mathbf{f}} - \mathbf{b}^T\hat{\mathbf{f}} + c$$

$$= -\frac{1}{4}\,\hat{\mathbf{f}}^T\mathbf{W}\hat{\mathbf{f}} - \frac{1}{4}\,\hat{\mathbf{f}}^T\mathbf{W}^T\hat{\mathbf{f}} - \mathbf{b}^T\hat{\mathbf{f}} + c$$

$$E = -\frac{1}{4}\,\hat{\mathbf{f}}^T\mathbf{W}\hat{\mathbf{f}} - \left(\frac{1}{4}\hat{\mathbf{f}}^T\mathbf{W}\hat{\mathbf{f}}\right)^T - \mathbf{b}^T\hat{\mathbf{f}} + c = -\frac{1}{2}\,\hat{\mathbf{f}}^T\mathbf{W}\hat{\mathbf{f}} - \mathbf{b}^T\hat{\mathbf{f}} + c \qquad (3.11)$$

From (3.11) we can see that the form of the energy functions given by using $\mathbf{W}$ and $\mathbf{W}^{\star}$ are identical. This means that if we assign neuron weights according to (3.10), the algorithm will converge.

The elements of $\mathbf{W}^{\star}$ can be considered an alternative approximation of (3.9) and are given by:

$$w_{ij}^{\star} = -\sum_{p=1}^{L} h_{pi}h_{pj} - \left(\frac{\lambda_i + \lambda_j}{2}\right)\sum_{p=1}^{L} d_{pi}d_{pj} \qquad (3.12)$$

However, since we have shown that both approximations produce identical energy functions, we will hence use the approximation given by (3.10). Next we need to show that approximating (3.9) as (3.10) little affects the quality of the restored images.

Equation (3.10) is much easier to calculate for each neuron since it only relies on the value of λ assigned to neuron i, and does not depend on the values of

λ assigned to other neurons in the neighborhood of neuron i. If we have a finite set of available values of λ, then by using (3.10) we will require only one weighing mask for each λ value in the set. If we instead use (3.9) we will have to recalculate the weights for each neuron in the network. This will, for obvious reasons, greatly degrade the performance of the network. The next question we must ask is how well (3.10) approximates (3.9) and when we will expect the approximation to break down.

Note that the concept of a *weighting mask* described in Chapter 2 used the fact that a neuron only has non-zero connections to other neurons within a certain neighborhood centered on the neuron. If a neuron lies in a region where all its neighbors were assigned the same value of λ, then (3.9) would be the same as the weighting mask produced by using a constant value of λ for all pixels in the image. This means that a value of λ can be found such that (3.9) can be accurately approximated by (3.10) and $w_{ij} = w_{ji}$ in the vicinity of this neuron.

If some of the neurons in the neighborhood of neuron i have different values of λ, but the difference between λ_i and λ_j is not great, then (3.10) is still a good approximation although $w_{ij} \neq w_{ji}$. However, we can expect that $w_{ij} \approx w_{ji}$.

If some of the neurons in the neighborhood of neuron i have very different values of λ, then the approximation will not hold. We cannot expect that $w_{ij} \approx w_{ji}$. This is not a big problem, however, since regions where neighboring neurons have greatly different values of λ must be edges and highly textured regions. The reasons for this will become apparent later. In such regions, noise is not readily noticeable and so any errors made by approximating (3.9) as (3.10) will be masked. Hence the approximation provides increased ease of use and only breaks down in those regions where the effects are least noticeable.

Since approximation (3.10) is much easier to implement than (3.9), this approximation will be used in the subsequent sections. The next question is how to determine the values of λ_i in (3.10) for each pixel in the image.

Let us investigate two methods for determining the regularization parameter. The first is an extension of the gradient-based method considered in the last chapter. In effect this method is motivated by the fact that since we used gradient descent to restore the image in Algorithm 2.3, then we may use gradient descent to find the optimal values of λ for each pixel. The second method is based on local statistics in the framework of an intelligent neural network.

3.3.2   The Gradient-Based Method

Since Algorithm 2.3 in the previous chapter used the principles of gradient descent to solve this problem, let us first look at how gradient descent might be used to select unique values of λ for each pixel.

Equations (2.7) and (2.8) indicate that differences in the regularization parameter affect only the weights of the neural network, not the initial values or the bias inputs.

Equation (2.12) gives the energy change resulting from a change of neuron state $\Delta \hat{f}_i$. Substituting (2.10) into (2.12) yields:

| (a) | (b) |

Figure 3.2: Original images

$$\Delta E_i = - \left[\sum_{j=1}^{L} w_{ij} \hat{f}_i + b_i \right] \Delta \hat{f}_i - \frac{1}{2} \, w_{ii} (\Delta \hat{f}_i)^2 \qquad (3.13)$$

It should be noted that $\sum_{j=1}^{L} w_{ij} \hat{f}_i + b_i$ is the input to each neuron and has the same sign as $\Delta \hat{f}_i$ if ΔE is to be negative. It is important to note that we are only considering the cases where ΔE is negative since when $\Delta E \geq 0$, there will be no pixel value update at pixel $\Delta \hat{f}_i$.

Definition 3.1: The greatest energy minimization (GEM) method for select-ing the constraint value for each pixel is defined as choosing for each pixel in each iteration the value of λ from a range of allowable λ values, $\lambda_a \leq \lambda \leq \lambda_b$, $\langle \lambda_a, \lambda_b \geq 0 \rangle \in \Re$, which best minimizes (3.13). [101, 111, 112]

Let us have a look at an example of applying this algorithm to real imagery. In this example, the images in Figure 3.2 were blurred using (3.5).

The number of choices of regularization parameter, R, was set to 5. The impulse response of the constraint matrix, $d(x, y)$, was as follows:

$$d(x, y) = \frac{1}{6} \begin{bmatrix} 1.0 & 1.0 & 1.0 & 1.0 & 1.0 \\ 1.0 & 4.0 & -3.0 & 4.0 & 1.0 \\ 1.0 & -3.0 & -20.0 & -3.0 & 1.0 \\ 1.0 & 4.0 & -3.0 & 4.0 & 1.0 \\ 1.0 & 1.0 & 1.0 & 1.0 & 1.0 \end{bmatrix}$$

Each image was degraded by the same PSF. For this experiment, a PSF of size $P = 5$ was used with $\sigma_x = \sigma_y = 2.0$. Various levels of white noise, with variances approximately equal to 4 and 18, were added to each degraded image.

Table 3.2 shows the SNR and LSMSE between the original and the degraded images, and images restored with a constant value of λ, and the greatest energy reduction technique for selecting λ, and the adaptive constraint technique of Kang and Katsaggelos [113].

Table 3.2: Statistics of degraded and restored images for various levels of noise

Noise	SNR(dB)			
	DI	NA	GER	KK
		Cat	*Image*	
4.24	12.61	13.30	12.55	11.38
18.30	12.25	12.81	11.02	10.83
		Flower	*Image*	
4.22	15.67	16.65	15.13	15.74
18.52	14.94	15.93	12.61	14.32
		LSMSE		
	DI	NA	GER	KK
		Cat	*Image*	
4.24	42.98	14.58	14.98	30.57
18.30	38.65	17.45	30.43	30.46
		Flower	*Image*	
4.22	20.53	4.35	9.74	6.56
18.52	19.35	5.94	33.54	10.88

Legend: DI = Degraded Image; NA = Non-Adaptively restored image; GER = Image restored using energy minimization method for constraint selection; KK = Image restored using the Kang and Katsaggelos algorithm.

Figures 3.3 and 3.4 show the degraded and restored images of the flower image, while Figure 3.5 shows the degraded and restored images of the cat image in the case of a noise variance of 18.30.

From Figures 3.3, 3.4 and 3.5, the images restored using the adaptive image restoration method appear clearer and more visually pleasing when compared to the images produced by the non-adaptive algorithm.

However, the results in Table 3.2 show that the algorithm is still less than perfect. Let us analyze this algorithm in more detail.

Analysis of the Gradient-Based Method

Based on the definition and other aforementioned assumptions, we present two theorems.

Theorem 3.1: If the GEM method is used to select a suitable λ value from a range of possible λ values, where $\lambda_a = 0$ and $\lambda_b = \infty$ in Definition 3.1, then λ_b will always be chosen.

Theorem 3.2: If the GEM method is used to select a suitable λ value, satisfying $\lambda_a \leq \lambda \leq \lambda_b$, where both λ_a and λ_b are finite, then either λ_a or λ_b will be chosen unless all available λ values would produce the same resultant decrease in energy.

Figure 3.3: Degraded and restored images for noise of variance 4.22

(a)   (b)

(c)   (d)

Figure 3.4: Degraded and restored images for noise of variance 18.52

Figure 3.5: Degraded and restored images for noise of variance 18.30

To prove the above theorems, we must rearrange (3.13). Expanding (3.13) using (3.1) we obtain:

$$\Delta E_i = \left[\sum_{j=1}^{L}\sum_{p=1}^{L} h_{pi}h_{pj}\hat{f}_j - b_i\right]\Delta\hat{f}_i + \lambda_i\Delta\hat{f}_i\sum_{j=1}^{L}\sum_{p=1}^{L} d_{pi}d_{pj}\hat{f}_j$$

$$+ \frac{1}{2}(\Delta\hat{f}_i)^2\sum_{p=1}^{L} h_{pi}^2 + \frac{1}{2}\lambda(\Delta\hat{f}_i)^2\sum_{p=1}^{L} d_{pi}^2$$

$$= A\Delta\hat{f}_i + \lambda_i B\Delta\hat{f}_i + C(\Delta\hat{f}_i)^2 + D\lambda_i(\Delta\hat{f}_i)^2 \qquad (3.14)$$

where $A = \sum_{j=1}^{L}\sum_{p=1}^{L} h_{pi}h_{pj}\hat{f}_j - b_i$, $B = \sum_{j=1}^{L}\sum_{p=1}^{L} d_{pi}d_{pj}\hat{f}_j$, $C = \frac{1}{2}\sum_{p=1}^{L} h_{pi}^2$, and $D = \frac{1}{2}\sum_{p=1}^{L} d_{pi}^2$.

It should be noted that $u_i = -A - \lambda B$ and

$$C + \lambda D = \frac{-w_{ii}}{2} \qquad (3.15)$$

Consider the $\Delta\hat{f}_i$ which maximizes the energy reduction for pixel i. Let this term be defined as $\Delta\acute{\hat{f}}_i$. To compute $\Delta\acute{\hat{f}}_i$, we differentiate (3.14) with respect to $\Delta\hat{f}_i$:

$$\frac{\delta\Delta E_i}{\delta\Delta\hat{f}_i} = A + \lambda_i B + 2C\Delta\hat{f}_i + 2D\Delta\lambda_i\Delta\hat{f}_i \qquad (3.16)$$

and set (3.16) to zero to obtain the optimal $\Delta\acute{\hat{f}}_i$.

$$\Delta\acute{\hat{f}}_i = \frac{-(A + \lambda_i B)}{2(C + \lambda_i D)} \qquad (3.17)$$

Rearrangement of (3.14) yields:

$$\Delta E_i = (A + \lambda_i B)\Delta\hat{f}_i + (C + \lambda_i D)(\Delta\hat{f}_i)^2 \qquad (3.18)$$

The substitution of (3.17) into (3.18) gives us:

$$\Delta\acute{E}_i = \frac{-(A + \lambda_i B)^2}{2(C + \lambda_i D)} + \frac{(A + \lambda_i B)^2}{4(C + \lambda_i D)} = \frac{-(A + \lambda_i B)^2}{4(C + \lambda_i D)} \qquad (3.19)$$

We can confirm from (3.17) that $\Delta\acute{\hat{f}}_i$ always has the same sign as u_i and that $\Delta\acute{E}_i$ is always negative for positive values of λ, as expected. The graph of $\Delta\acute{E}_i$ versus λ_i is sketched in Figure 3.6.

We can see from (3.19) that for all acceptable values of ΔE_i, $\lim_{\lambda_i\to\infty}\Delta\acute{E}_i = -\infty$. This proves Theorem 3.1 and implies that a great energy reduction can always be obtained by choosing a very large value of λ_i. If we allow energy

Figure 3.6: Graph of maximum energy reduction versus constraint value

considerations alone to dictate the choice of λ_i, the results would be poor. An acceptable range of λ values must be set. Since C and D are always positive, and $\Delta\acute{E}_i$ is always negative, (3.19) has a maximum at $\lambda_i = -\frac{A}{B}$.

Since (3.19) is a convex function when $\Delta\acute{E}_i < 0$, then we observe that if the set of acceptable λ values is finite and constrained between λ_a and λ_b ($\lambda_a < \lambda_b$), then (3.19) indicates that under most circumstances, the best energy reduction will be obtained by choosing either λ_a or λ_b, and not any intermediate values of λ. Only under the conditions of $\frac{-A^2}{4C} \approx \frac{-B^2}{4D} \approx 0$ will numerical errors possibly enable the choice of intermediate values of λ. This proves Theorem 3.2.

From this analysis, a number of conclusions can be drawn. When A is low and B is high, the largest λ_i value gives the best energy minimization; however, when A is high and B is low, the smallest value of λ_i produces the best energy minimization.

The next question is which λ values will be chosen in low and high variance regions of the image to maximize the energy reduction. To clearly answer this question, it is necessary to examine how the factors A, B, C and D vary as a function of image statistics.

Let's look at each of the factors in turn.

Factor A

Factor A is given by:

$$A = \sum_{j=1}^{L} \sum_{p=1}^{L} h_{pi} h_{pj} \hat{f}_j - b_i = K - b_i$$

where $K = \sum_{j=1}^{L} \sum_{p=1}^{L} h_{pi} h_{pj} \hat{f}_j = [\mathbf{H}^T \mathbf{H} \hat{f}]_i$.

The factor K is the value of pixel $\hat{f}_i$ after a double application of the degrading PSF to the image estimate. Using the approximate formula for $\mathbf{H}$, then $\mathbf{H}$ is symmetric, $\mathbf{H}^T = \mathbf{H}$. Note that since $b_i = \sum_{p=1}^{L} h_{pi} g_p$, b_i is the value of pixel $\hat{f}_i$ after a single application of the degrading PSF to the degraded image, which is itself the result of applying the degrading PSF to the original image. Therefore in effect:

$$b_i = \sum_{p=1}^{L} \sum_{j=1}^{L} h_{pi} h_{pj} f_j = [\mathbf{H}^T \mathbf{H} \mathbf{f}]_i \qquad (3.20)$$

where $\mathbf{f}$ is the estimate of the original when noise is not considered. The algorithm would eventually return $\mathbf{f}$ as the restored image when $\lambda = 0$. Hence we obtain:

$$\mathbf{A}_k = \mathbf{H}^T \mathbf{H} \hat{\mathbf{f}} - \mathbf{H}^T \mathbf{H} \mathbf{f} = \mathbf{H}^T \mathbf{H} (\hat{\mathbf{f}} - \mathbf{f}) \qquad (3.21)$$

where $\mathbf{A}_k$ is the column vector whose element i is the value of factor A computed at neuron i during the kth iteration of the algorithm.

Considering (3.21), it is obvious that the entries in $\mathbf{A}_k$ may be positive or negative and will approach zero as the image estimate approaches the original image in the event of no noise. We can expect $\mathbf{A}_k$ to have its greatest values in the initial iteration of the algorithm.

If the initial image estimate is $\hat{\mathbf{f}} = \mathbf{g} = \mathbf{H} \mathbf{f}$, then:

$$\mathbf{A}_1 = \mathbf{H}^T \mathbf{H} (\mathbf{H} \mathbf{f} - \mathbf{f}) = \mathbf{H}^T \mathbf{H} (\mathbf{H} - \mathbf{I}) \mathbf{f} \qquad (3.22)$$

As long as the additive noise is not too severe, the double application of the low-pass degrading PSF given by $\mathbf{H}$ will remove most of the noise in the smooth regions of the image. Since $\mathbf{H} - \mathbf{I}$ is a high-pass filter, factor A would tend to be large in high variance regions and small in low variance regions of the image.

Factor B

For factor B we have $B = \sum_{j=1}^{L} \sum_{p=1}^{L} d_{pi} d_{pj} \hat{f}_j$, which can be rewritten as:

$$B = [\mathbf{D}^T \mathbf{D} \hat{\mathbf{f}}]_i \qquad (3.23)$$

Since $\mathbf{D}^T = \mathbf{D}$, factor B is hence the value of pixel $\hat{f}_i$ after a high-pass filter has been applied twice to the image estimate. On edges and high texture regions, we would expect the magnitude of factor B to be large. However, since the high-pass filter is applied twice, noise and very sharp edges would produce a larger magnitude of B than more gradual edges in the image. It is important to note that B may be positive or negative depending on whether the value of $\hat{f}_i$ is higher or lower than the mean of its neighbors; however, B will tend to zero in low variance regions of the image.

Factors C and D

Factors C and D are given by $C = \frac{1}{2} \sum_{p=1}^{L} h_{pi}^2$ and $D = \frac{1}{2} \sum_{p=1}^{L} d_{pi}^2$. Both these factors are constant. C will always be quite small and in the case of a degrading PSF modeled by a 5 by 5 Gaussian blur of standard deviation 2.0 in both x and y directions, $C = 0.28$. Factor D depends upon the type of high-pass filter chosen and typical values are of the order of 50 to 500.

Observation 1: In high texture regions of the image or near edges, both A and B can be expected to be large. However, the contribution from factor B tends to be less significant for two reasons.

1. The factor B in (3.19) is multiplied by λ_i which is always $\ll 1$.

2. The factor B corresponding to a double application of a high-pass filter suggests that in the presence of gradual edges, B will not be very large. Since noise is added to the image after blurring, the presence of noise will contribute most to the value of B.

Therefore we can expect that $\frac{-A^2}{4C} < \frac{-\lambda_b B^2}{4D}$, and so the best energy reduction would be obtained by choosing the lowest value of λ available.

Observation 2: In low texture regions of the image, A can be expected to be small. Factor B may also be small. However, the presence of noise in the initial estimate should make the effect of B on (3.19) override that of A. Hence in low texture regions of the image, choosing the largest available λ_i produces the best energy minimization.

This analysis has been verified in practice. Figure 3.7 shows the selection of λ values during the first iteration for an image degraded by a 5 by 5 Gaussian PSF of standard deviation 2.0, with additive noise of variance 18.49. The darker regions in Figure 3.7 denote the selection of larger λ values, and the lighter regions denote the selection of smaller λ values.

From Figure 3.7 we can clearly see that to a great extent, only the lowest or the highest value of λ is chosen by the gradient descent method. Although mathematically elegant, using gradient descent to select the regularization parameter does not agree with biological vision. In addition, this method does not allow us to use a wide range of λ values to restore the image. Only two λ values are used. This effectively divides the image into low and high variance regions.

3.3.3   Local Statistics Analysis

The above analysis shows us that gradient descent is not an optimal approach for choosing λ_i in adaptive regularization. However, it leads us to the observation that in low texture regions, a high value of λ_i results in a visually pleasing result, while in high texture regions, a low λ_i value works best. These observations encourage us to use local statistics, since these are a good measure of the image roughness to locally determine the λ_i value. Mathematically:

Figure 3.7: The λ values selected for each pixel during the initial iteration of the energy minimization-based algorithm for a typical image

$$\lambda_i = Y(S_i) \tag{3.24}$$

where $Y(S_i)$ is a function of the local image statistics S_i at $\hat{f}_i$.

Since S_i has almost exactly the same value for all pixels, $\hat{f}_i$, in a statistically homogeneous area, the λ_i value is also almost exactly the same. Therefore the structure of the processing model can be further modified such that, instead of assigning each pixel a different λ_i, we assign each statistically homogeneous area a λ_k.

Assume that there are K homogenous areas. By first properly rearranging the pixels in such a way that the pixels in a homogenous area are consecutively indexed in $\hat{\mathbf{f}}$ to form a new vector $\hat{\mathbf{f}}^\star$, equation (3.7) can be rewritten as:

$$E = \frac{1}{2}\{\|\mathbf{g}^\star - \mathbf{H}^\star\hat{\mathbf{f}}^\star\|^2 + \hat{f}^\star(\mathbf{D}^{\star T}\mathbf{\Lambda}^\star\mathbf{D}^\star)\hat{f}^\star\} \tag{3.25}$$

where $\hat{\mathbf{f}}^{\star T} = [\mathbf{f}_1^T, \ldots, \mathbf{f}_K^T]$ with $\mathbf{f}_k$ being the vector consisting of the pixels in the kth homogenous area,

$$\mathbf{H}^\star = \begin{bmatrix} \mathbf{H}_1 \\ \ldots \\ \mathbf{H}_K \end{bmatrix}, \mathbf{\Lambda}^\star = \begin{bmatrix} \mathbf{\Lambda}_1 & 0 & 0 \\ 0 & \ldots & 0 \\ 0 & 0 & \mathbf{\Lambda}_K \end{bmatrix}, \mathbf{D}^\star = \begin{bmatrix} \mathbf{D}_1 \\ \ldots \\ \mathbf{D}_K \end{bmatrix}, \mathbf{g}^\star = \begin{bmatrix} \mathbf{g}_1 \\ \ldots \\ \mathbf{g}_K \end{bmatrix} \cdot \tag{3.26}$$

with $\mathbf{H}_k$, $\mathbf{\Lambda}_k$, $\mathbf{D}_k$ and $\mathbf{g}_k$ being the submatrices (vectors) of $\mathbf{H}^\star, \mathbf{\Lambda}^\star, \mathbf{D}^\star$ and $\mathbf{g}^\star$, corresponding to $\mathbf{f}_k$, and $\mathbf{\Lambda}_k = \lambda_k \mathbf{I}$, with $\mathbf{I}$ being the identity matrix. Define $\mathbf{H}_k = [\mathbf{H}_{k l}, \ldots, \mathbf{H}_{kK}]$ and $\mathbf{D}_k = [\mathbf{D}_{kl}, \ldots, \mathbf{D}_{kK}]$, $k = 1, 2, \ldots, K$; then mathematical manipulation of (3.25) leads to:

$$E = -\tfrac{1}{2} \sum_{k=1}^{K} \{ [\hat{\mathbf{f}^\star}^T (\mathbf{H}_k^T \mathbf{H}_k + \lambda_k \mathbf{D}_k^T \mathbf{D}_k) \hat{\mathbf{f}^\star} + 2\mathbf{g}_k^T \mathbf{H}_k \hat{\mathbf{f}^\star}] \} + \|\mathbf{g}^\star\|^2$$

$$E = -\tfrac{1}{2} \sum_{k=1}^{K} \{ \mathbf{f}_k^T (\mathbf{H}_{kk}^T \mathbf{H}_{kk} + \lambda_k \mathbf{D}_{kk}^T \mathbf{D}_{kk}) \mathbf{f}_k$$

$$+ \sum_{l=1, l\neq k}^{K} \sum_{m=1, m\neq k}^{K} \mathbf{f}_l^T (\mathbf{H}_{kl}^T \mathbf{H}_{km} + \lambda_k \mathbf{D}_{kl}^T \mathbf{D}_{km}) \mathbf{f}_m + 2\mathbf{g}_k^T \mathbf{H}_k \hat{\mathbf{f}^\star} \} + \|\mathbf{g}^\star\|^2 \quad (3.27)$$

Apparently, $\mathbf{H}_{kk}^T \mathbf{H}_{kk} + \lambda_k \mathbf{D}_{kk}^T \mathbf{D}_{kk}$ represents the intra-connections within area k, and $\mathbf{H}_{kl}^T \mathbf{H}_{km} + \lambda_k \mathbf{D}_{kl}^T \mathbf{D}_{km}$ represents the inter-area contributions from areas l and m to area k. Equation (3.27) is the extension of a biologically motivated neural network: The network of networks (NoN) [114]. The significance of this mapping is as follows:

1. The human visual system pays little attention to individual pixels in an image. Instead it looks for areas of similar pixels in a statistically homogeneous sense, and it is more sensitive to edges [85].

2. The NoN is a computational model imitating a simplified representation of the human cortex, or part of it, the biological visual processing machine. The neurons in the same cluster are similar to one another.

3. By representing a pixel with a neuron, a homogenous image area is mapped to a cluster of neurons which are similar to one another. Local statistics is a good criterion to measure the level of similarity.

Therefore, using a NoN with a statistical criterion to adaptively determine the λ value and in turn the processing architecture may potentially simulate some aspects of human perception in recovering genuine information lost in recording.

Now the important issue is selecting the λ_i for each homogeneous area. In this section, we present a curve fitting approach [101, 102, 115, 116]. In this approach, the largest and the smallest values of S_i, S_{imax} and S_{imin}, are identified and the corresponding λ_{imax} and λ_{imin} are determined.

Experiments performed by the authors found that if the variance in a region gradually increases by equal steps, then the change in variance level is much more noticeable when the overall variance levels are low rather than when the overall variance levels are high. Humans are less able to discern an increase in noise levels in high variance regions than they are for low variance regions. A log-linear relationship is therefore suggested:

$$\lambda_i = a \, \log(S_i) + b \quad (3.28)$$

In Figure 3.8, six images with a constant level of pixel intensity are superimposed by noise of increasing variance. The increasing level of noise between each

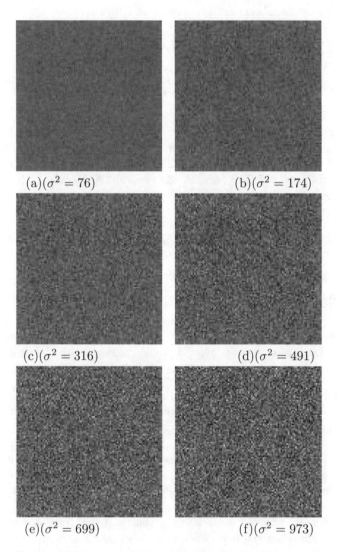

Figure 3.8: Images with varying levels of additive noise

image and its neighbor on the right is readily noticeable for the first three images when variance levels are low, compared to the last three when variance levels are high. The difference in apparent noise level between the last two images is the weakest of the entire set of images, despite the fact that the increase in variance between these two images is the greatest of the set.

The above observations lend weight to the use of constant λ values for each iteration based on those during the first iteration. Since the λ selected during the first iteration can often produce visually pleasing results (large λ in the background and small λ on the edges), these can be held constant throughout the restoration procedure. The λ values may not act as favorably for later iterations in the algorithm. For example, the local variance levels in the areas surrounding edges in the image may increase during the restoration due to ripple effects from nearby edges such that they become large enough to cause a large value of λ to be applied to edges of the image, hence producing the opposite effect to that which we are attempting to achieve. Let us look now at some examples of this algorithm when applied to real imagery.

This example is an extension of the example given in Section 3.3.2. As with the example in Section 3.3.2, a PSF of size $P = 5$ was used with $\sigma_x = \sigma_y = 2.0$. Various levels of white noise, with variances approximately equal to 4 and 18, were added to each degraded image.

Table 3.3 is an extension of Table 3.2 and shows the SNR and LSMSE between the original and the degraded images, and images restored with a constant value of λ, the greatest energy reduction technique for selecting λ, the adaptive constraint technique of Kang and Katsaggelos [113], and with an adaptive λ value based on local image variance and (3.28). Figure 3.9 shows regularization parameters used in the experiment plotted against the level of local variance. The values of λ were associated with variance thresholds in a way consistent with (3.28).

Figures 3.10 and 3.11 show the degraded and restored images of the flower image, while Figure 3.12 shows the degraded and restored images of the cat image in the case of a noise variance of 18.30.

From Figures 3.10, 3.11 and 3.12, the images restored using the adaptive image restoration methods appear clearer and more visually pleasing despite a slight decrease in SNR when compared to the images produced by the non-adaptive algorithm. Table 3.3 indicates that using variance as a criteria to choose the value of regularization parameter produces images with an improved LSMSE as noise levels increase.

Figure 3.9: Constraint values chosen versus variance

Table 3.3: Statistics of degraded and restored images for various levels of noise

Noise	SNR(dB)				
	DI	NA	GER	KK	VB
		Cat		Image	
4.24	12.61	13.30	12.55	11.38	13.19
18.30	12.25	12.81	11.02	10.83	12.58
		Flower		Image	
4.22	15.67	16.65	15.13	15.74	16.95
18.52	14.94	15.93	12.61	14.32	15.82
			LSMSE		
	DI	NA	GER	KK	VB
		Cat		Image	
4.24	42.98	14.58	14.98	30.57	13.51
18.30	38.65	17.45	30.43	30.46	15.79
		Flower		Image	
4.22	20.53	4.35	9.74	6.56	2.53
18.52	19.35	5.94	33.54	10.88	3.74

Legend: DI = Degraded Image; NA = Non-Adaptively restored image; GER = Image restored using energy minimization method for constraint selection; KK = Image restored using the Kang and Katsaggelos algorithm; VB = image restored using Variance Based constraint selection.

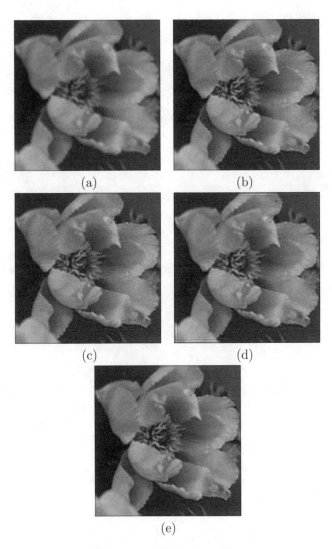

Figure 3.10: Degraded and restored images for noise of variance 4.22

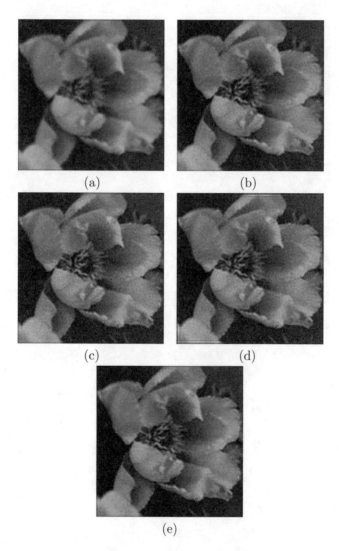

Figure 3.11: Degraded and restored images for noise of variance 18.52

Figure 3.12: Degraded and restored images for noise of variance 18.30

3.4   Correcting Spatially Variant Distortion Using Adaptive Constraints

In the previous section, the weights of the neural network were varied spatially in order to implement the adaptive regularization parameter. A similar concept was considered in Section 3.2 to handle a simple form of spatially variant distortion. In Section 3.2, we saw how multiple sets of weights could be used to restore an image degraded by a cyclic variation of Gaussian PSFs, of different standard deviations. In this section, we will examine the integration of the spatially variant restoration technique of Section 3.2 with the adaptive constraint technique of Section 3.3.

The method employed in Section 3.2 to handle spatially variant distortion was to precompute a number of sets of weights to handle the various PSFs degrading the image. If the spatial variation of the PSFs was known, the correct set of weights could be chosen to restore the image accurately. However, a fixed λ was used in Section 3.2. To implement a spatially variant restoration technique with an adaptive constraint parameter, we produce as many sets of weights as needed to handle the space variant degradation and every possible choice of regularization parameter.

Let us consider as an example, the same type of distortion considered in Section 3.2. Restating the discussion in Section 3.2, the cyclic spatially variant distortion we consider is that obtained by the use of V PSFs $h_0(x, y), \ldots, h_{V-1}(x, y)$. The pixels in any one row of the image are acted upon by the same PSF; however, the PSF applied to each row is varied cyclically through the sequence:

$$S_H = \{h_0(x, y), h_1(x, y), \ldots, h_{V-1}(x, y), h_{V-2}(x, y), \ldots, h_1(x, y)\} \qquad (3.29)$$

The sequence S_H has a period of $2V - 2$, and hence $2V - 2$ unique sets of weights are required to restore the image when the same regularization parameter is chosen for every pixel. When additional sets of weights are created to implement an adaptive regularization parameter scheme, the following analysis applies:

Taking R as the number of choices of regularization parameter to be used in the restoration, then the sets of weights to be used to restore a row blurred by the ith element of sequence S_H form the set:

$$\mathbf{WSM}_i = \{\mathbf{Wm}_0^i, \ldots, \mathbf{Wm}_{R-1}^i\} \qquad (3.30)$$

where $\mathbf{Wm}_j^i$ is the set of weights required to restore a row degraded by PSF $[S_H]_i$ using the jth choice of regularization parameter being considered. The restoration problem becomes a problem of selecting the correct set of weights from the super set of:

$$\mathbf{WMA} = \{\mathbf{WSM}_0, \ldots, \mathbf{WSM}_{2V-3}\} \qquad (3.31)$$

Restoration of the image is accomplished by selecting the relevant set $\mathbf{WSM}_i$, based on the row coordinate of the current pixel being examined, then selecting the optimal $\mathbf{Wm}_j^i$ within that set based on the selection schemes described in Section 3.3 [101]. Let's look now at an example of implementing the spatially variant restoration scheme with an adaptive regularization parameter as described above. An image was created using a cyclic variation of 7 by 7 Gaussian PSFs. Using the analysis in Section 3.2, V was set to be 4. As before, Table 3.1 details the degrading PSFs used to blur the image as per equation (3.29). Noise of variance 4.28 was added to the blurred image. The degraded image is shown in Figure 3.13a. First, the image was restored using the non-adaptive spatially invariant approach. A regularization parameter value of 0.0007 was used and the spatially variant distortion was approximated as a spatially invariant distortion by using a 7 by 7 gaussian PSF of standard deviation 2.55. This image is shown in Figure 3.13b.

The image was then restored using a non-adaptive spatially variant restoration method, with a regularization parameter value of 0.0007. This image is shown in Figure 3.13c. In the final approach, the degraded image was restored using a spatially variant, adaptive constraint method, with the variance based λ value selection algorithm. The regularization value was selected in the same manner as the example in the previous section. This image is shown in Figure 3.13d. The statistics are summarized in Table 3.4.

We can see from Figure 3.13d that the Space Variant Adaptive restoration approach has produced a visually more pleasing result with improved values of

(a)   (b)

(c)   (d)

Figure 3.13: Results of experiments on an image degraded by spatially variant distortion

Table 3.4: Results of spatially variant experiment

Image	SNR (dB)	LSMSE
Degraded	13.94	36.01
Space Invariant Non-adaptive	15.03	6.38
Space Variant Non-adaptive	15.87	6.04
Space Variant Adaptive	16.01	4.84

SNR and LSMSE compared to the other methods.

3.5   Semi-Blind Restoration Using Adaptive Constraints

The previous sections have assumed that the nature of the degrading PSF is known. In practical circumstances, however, this may not be the case. We may have little or no knowledge of the degrading PSF in the case of a spatially invariant distortion or may not know the exact nature of the space variance in a case of spatially variant distortion. In addition, our estimate of the degrading PSF may be corrupted by noise. Restoring an image with incomplete knowledge of the nature of the degradation is called *semi-blind deconvolution*. Figure 3.14 shows the effects of using an incorrect PSF estimate during the restoration procedure.

Figure 3.14a is the original flower image. Figure 3.14b is the flower image degraded by a 5 by 5 Gaussian blur of standard deviation 2.0 with additive noise of variance 4.22. Figure 3.14c shows the image restored using the parameters of a 7 by 7 Gaussian blur of standard deviation 3.0, and Figure 3.14d shows the image restored using the parameters of a 3 by 3 Gaussian blur of standard deviation 1.0. In Figure 3.14c, the degrading PSF estimate is too strong and so the image can be said to have been *over restored*. Although the edges and high variance regions in Figure 3.14c appear sharp, ringing effects are apparent in the smooth (low variance) regions of the image. In Figure 3.14d, on the other hand, the degrading PSF estimate is too weak and so the image can be said to have been *under restored*. In the smooth (low variance) regions of Figure 3.14d, we see no artifacts; however, the edges and high variance regions of Figure 3.14d appear blurry.

This would indicate that a too strong PSF estimate produces problems in low variance regions whereas a too weak PSF estimate produces problems in high variance regions. Figure 3.14d appears similar to the result obtained when a large value of λ is used during the restoration process. In fact, since λ controls the degree of smoothness in the solution, the effect of increasing λ is similar to performing the restoration with a weaker PSF estimate. In the adaptive λ image restoration methods described in the previous sections, a small value of λ is used in high variance regions and a larger value of λ is used in low variance regions.

Figure 3.14: Degraded image restored with various PSF estimates

This implies that the adaptive constraint restoration method described above may have an application in semi-blind image restoration [102]. An example should illustrate this.

Assume that an image has suffered a spatially variant distortion and only an estimate of the average PSF is available. By using the adaptive constraint restoration methods developed in Section 3.3 we can consider that each different value of λ used in the methods approximates a different strength PSF estimate and hence we effectively have a range of available PSF estimates to choose from. In the high variance regions of the image, the λ value is small, and this is similar to selecting the strongest PSF estimate. This is the preferable choice since over-restoring the high variance regions of the image is less visually disturbing than under-restoring them. To reduce the effects of ripples in the low variance regions of the image a weak PSF estimate is preferable. This is automatically achieved using the adaptive constraint methods described in Section 3.3 since these methods select the largest value of λ for low variance regions. As mentioned above, larger values of λ produce results similar to using weaker PSF estimates. From this we can see that the adaptive λ restoration method we developed in Section 3.3 can be used to compensate for inaccuracies in the estimate of degrading PSF. Let's expand the example in Section 3.4 to examine this effect.

An image was created using a cyclic variation of 7 by 7 Gaussian PSFs. Using the analysis in Section 3.2, V was set to be 4. Table 3.1 details the degrading PSFs used to blur the image as per equation (3.29).

Noise of variance 4.28 was added to the blurred image. The degraded image is shown in Figure 3.15a. This image was restored using four techniques. The first technique was a non-adaptive spatially invariant approach. As in the previous section, a regularization parameter value of 0.0007 was used and the spatially variant distortion was approximated as a spatially invariant distortion by using a 7 by 7 Gaussian PSF of standard deviation 2.55. This image is shown in Figure 3.15b. The second technique was an adaptive spatially invariant approach. The same approximation of the PSF was in the previous experiment was used; however, the regularization parameter was varied for each pixel using the same local variance levels and λ values as in the low noise example in Section 3.3.3. This image is shown in Figure 3.15c.

The image was then restored using a non-adaptive spatially variant restoration method, with a regularization parameter value of 0.0007. This image is shown in Figure 3.15d. In the final approach, the degraded image was restored using a spatially variant, adaptive constraint method, with the variance based λ value selection algorithm. The regularization value was selected in the same manner as the example in Section 3.3.3. This image is shown in Figure 3.15e. The statistics are summarized in Table 3.5. It is interesting to note that by using an adaptive approach we can compensate for a lack of knowledge regarding the degrading PSF. In the cases where the degraded images are restored by the space invariant approach, we can see that using the adaptive technique produces a much clearer image with a lower LSMSE although the cyclic blurring effect is marginally visible. In all cases, using the adaptive approach produces a clearer image with a lower LSMSE.

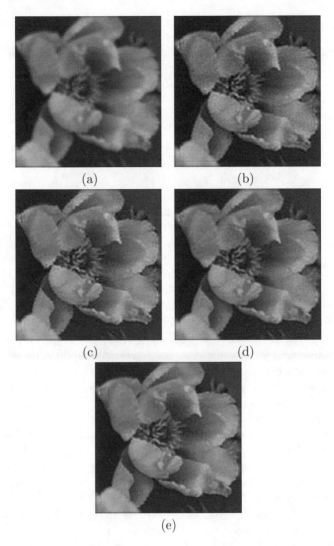

Figure 3.15: Results of experiments on an image degraded by spatially variant distortion

Table 3.5: Results of spatially variant experiment

Image	SNR (dB)	LSMSE
Degraded	13.94	36.01
Space Invariant Non-adaptive	15.03	6.38
Space Invariant Adaptive	14.92	5.45
Space Variant Non-adaptive	15.87	6.04
Space Variant Adaptive	16.01	4.84

3.6   Implementation Considerations

A major practical consideration is the number of choices of regularization value in the local statistics-based adaptive constraint restoration algorithm in Section 3.3. Taking R as the number of choices of regularization parameter to be used in the restoration, one would expect that a large value of R will give a better restoration quality; however, large values of R would slow the constraint precomputing stage of the algorithm and waste memory. It is desirable to have as low a value of R as possible. In practice setting $R = 3$ is usually sufficient, giving the algorithm the choice of doing nothing or selecting one of three constraint values for regions of high, medium or low texture levels.

For the numerical results in this chapter we associated each set of weights with a range of variance values computed in a certain neighborhood of size A by A of the current pixel. A variance threshold, κ, was set, below which the pixel being examined would not be updated. This is because for extremely low variance regions of an image blurring may not be noticeable; in this case restoration can only serve to enhance noise and waste time. This technique can yield improved results; however, the level of variance threshold and whether or not a variance threshold is used at all depend on the image being restored and the degrading function. The variance constraint selection method has the advantage that it does not require the current pixel to be acted upon by each possible set of weights to compute the regularization parameter required. Another advantage to the variance method is the ability to fine-tune the variance thresholds, κ, or A, the area size, to suit a particular type of image being examined.

When we consider using variance to determine the regularization parameter, we expect that precomputation of the parameter, based on the degraded image statistics would produce similar results to computing the parameter in each iteration. High variance regions in the degraded image should remain high variance in the restored image, and low variance regions should likewise remain low variance; hence by this argument the chosen values of regularization parameter in the first iteration should remain approximately the same throughout the restoration procedure. This also has the added advantage that the algorithm is guaranteed to converge according to the analysis in Chapter 2.

In practice, precalculating the optimal set of weights for each pixel has further advantages:

- During restoration, the estimate may converge on the solution smoothly or with varying degrees of oscillation around the final value. During an oscillatory restoration, the image statistics may change in unpredictable ways, causing the regularization parameter chosen from iteration to iteration to vary also. This may not result in an optimal choice of regularization value during any one iteration or an optimal average value of regularization parameter.

- Precomputation of the regularization parameter results in a faster restoration than that obtained by computation of the regularization parameter during each iteration. In fact, by precomputing the regularization values for each pixel, the adaptive constraint algorithm takes only slightly more time than a non-adaptive algorithm for large images, as will be shown in the next section.

3.7 More Numerical Examples

Let us look at two more examples of how the algorithms we have presented in this chapter perform on real imagery. The first example examines the processing efficiency. Following this, a practical example of the use of this method will be given. As with previous examples, the images will be compared by measuring their Signal to Noise Ratios (SNR) and LSMSEs (LSMSE is as defined in Chapter 1).

3.7.1 Efficiency

In this example, the time taken to restore an image was compared among the four different cases from the space variant example in Section 3.5. Each algorithm was run three times on a SUN Ultra 1 workstation, the average results for each algorithm are shown in Table 3.6.

It is worth noting that the execution times of the adaptive regularization parameter algorithms are similar to the execution times of the non-adaptive algorithms. The adaptive algorithms require the calculation of five times the number of weighting masks as the non-adaptive algorithms, and also require pre-restoration calculations of local variance to precompute the regularization parameter for each pixel. The reason for the similar execution times is that the time lost through setting up the adaptive parameters of the network is offset by the time gained through the non-adjustment of pixels in low variance regions of the image. The most important fact that we can observe from Table 3.6 is that the time difference required for a fully adaptive, spatially variant restoration is only double the time required for the much simpler non-adaptive spatially invariant restoration. This is much faster than any previously reported methods. In fact as the size of the image increases we would expect the time required for adaptive spatially variant restoration to approach the time required for non-adaptive, spatially invariant restoration. This is because the extra time required for the adaptive spatially variant restoration is primarily taken up by the initial

Table 3.6: Algorithm run times

Algorithm	Time (CPU seconds)
Non-Adaptive Spatially Invariant	259
Non-Adaptive Spatially Variant	370
Adaptive Spatially Invariant	206
Adaptive Spatially Variant	667

extra weighting mask creation. The quality and speed of the adaptive spatially invariant approximation method offers a promising alternative to the traditional semi-blind deconvolution methods, especially when the PSF is space variant.

3.7.2   An Application Example

The above neural network algorithm was applied to the problem of restoring images with an unknown level of blur. Images were supplied to us showing chromosomes in a solution imaged by a high powered optical microscope. Limitations of the optical system had blurred the images of the chromosomes. To extract further information about the chromosomes in the images, the above neural network algorithm was used to enhance the images. Figure 3.16a shows one of the original degraded images. Figure 3.16b shows Figure 3.16a restored using the non-adaptive algorithm. Note that the restored image is sharper than the original image, but some noise has also been amplified as well as some ringing effects. Figure 3.16c shows Figure 3.16a restored using the adaptive algorithm. The level of sharpness is comparable to the results of the non-adaptive approach; however, the level of background noise and ringing effects have been greatly reduced. In this case the adaptive restoration algorithm has been successful at enhancing the detail present in the image.

(a)

(b)

(c)

Figure 3.16: Degraded and restored chromosome images

3.8   Adaptive Constraint Extension of the Lagrange Model

Section 3.3 developed the adaptive constraint problem by considering (3.6) as a form of simple penalty function. However, (3.6) may also be considered in terms of the theory of Lagrange [117]. We will see in this section the consequences of this analysis. First we will expand the restoration problem in terms of the theory of Karush-Kuhn-Tucker (a generalization of the theory of Lagrange) and then show what solutions result. Finally we will investigate what conditions on the problem need to be imposed for the analysis to be valid.

3.8.1   Problem Formulation

Consider the image degradation model:

$$\mathbf{g} = \mathbf{Hf} + \mathbf{n} \tag{3.32}$$

where $\mathbf{g}$ and $\mathbf{f}$ are L-dimensional lexicographically organized degraded and original images, respectively, $\mathbf{n}$ is an L-dimensional additive noise vector and $\mathbf{H}$ is a convolutional operator. Assume that the image $\mathbf{f}$ can be divided into K regions with common statistical properties, where each pixel belongs to one region and only one region. The regions are such that for any pixel in region i, any other pixel in region i may be reached by a path consisting only of pixels belonging to region i.

Let $\{i\}$ denote the set of lexicographic indices of pixels in region i. Let's define an operator that extracts only those pixels belonging to region i.

Define $\mathbf{K}_i \in \Re^{L \times L}, 1 \le i \le K$ where

$$[\mathbf{K}_i]_{jl} = \begin{cases} 1, & \text{if } j = l \text{ and } j, l \in \{i\} \\ 0, & \text{if } j \notin \{i\} \text{ or } l \notin \{i\} \\ 0, & \text{if } j \ne l \end{cases} \tag{3.33}$$

Hence $\mathbf{K}_i \mathbf{f}$ is a vector containing only pixels belonging to region i, that is:

$$[\mathbf{K}_i]_{jl} = \begin{cases} f_j, & \text{if } j \in \{i\} \\ 0, & \text{otherwise} \end{cases}$$

Note that $\mathbf{K}_i^T = \mathbf{K}_i \mathbf{K}_i = \mathbf{K}_i$, due to the simple structure of $\mathbf{K}_i$.

The problem we are considering is that given $\mathbf{g}$, and $\mathbf{H}$, we wish to recover $\mathbf{f}$. Assume that using the degraded image, $\mathbf{g}$, we can reasonably partition the expected result $\mathbf{f}$ into the K regions described above. The pixels in each region have similar statistical properties.

Such a problem is ill-posed and so may have an infinite number of solutions. For this reason, extra constraints will be required. One reasonable constraint is to require that the value of every pixel in the image be positive. Consider a matrix $\mathbf{D} \in \Re^{L \times L}$ designed to act as a high-pass operator. $\mathbf{Df}$ is therefore the high-pass component of the estimated image. To reduce the effects of noise, we can constrain $\|\mathbf{Df}\|^2$ to be equal to a preset constant. If the constant is chosen too low, the image will be blurred, and if the constant is chosen too high, the image will contain unacceptable levels of noise. However, the best value of the constant to use for each region of the image may not be same as every other region. A better approach is to constrain $\|\mathbf{Df}\|^2$ to be equal to different constants in different parts of the image.

This problem can then be stated as:

$$\mathbf{f} = \arg \min_{\mathbf{f}} \|\mathbf{g} - \mathbf{Hf}\|^2 \text{ subject to } \mathbf{h}(\mathbf{f}) = \mathbf{0} \text{ and } \mathbf{p}(\mathbf{f}) \le \mathbf{0}$$

where $\mathbf{h}(\mathbf{f}) \in \Re^K, h_i(\mathbf{f}) = \|\mathbf{K}_i\mathbf{Df}\|^2 - \sigma_i^2, 1 \le i \le K$ and $\mathbf{p}(\mathbf{f}) \in \Re^L, \mathbf{p}(\mathbf{f}) = -\mathbf{f}, p_j(\mathbf{f}) = -f_j, 1 \le j \le L$.

Note that $\mathbf{p}$ is the set of positivity constraints, while $\mathbf{h}$ comprises the constraints on the norm of the high frequency components of each region.

Let us look now at how we can solve this equation. The following definition and theory will be useful.

Definition 3.2: Let $\mathbf{f}^\star$ satisfy $\mathbf{h}(\mathbf{f}^\star) = \mathbf{0}, \mathbf{p}(\mathbf{f}^\star) \le \mathbf{0}$ *and let* $J\{\mathbf{f}^\star\}$ *be the index set of active inequality constraints, that is,* $J\{\mathbf{f}^\star\} = \{j; p_j(\mathbf{f}^\star) = 0\}$.

Then $\mathbf{f}^\star$ *is a regular point if* $\nabla h_i(\mathbf{f}^\star), \nabla p_j(\mathbf{f}^\star)$ *are linearly independent* $\forall i, j; 1 \le i \le K, j \in J\{\mathbf{f}^\star\}$. [117]

Theorem of Karush-Kuhn-Tucker:

Let $E, \mathbf{h}, \mathbf{p} \in C^1$. *Let* $\mathbf{f}^\star$ *be a regular point and a local minimizer for the problem of minimizing* E *subject to* $\mathbf{h}(\mathbf{f}^\star) = \mathbf{0}$ *and* $\mathbf{p}(\mathbf{f}^\star) \le \mathbf{0}$.

Then there exists $\lambda \in \Re^K, \mu \in \Re^L$ *such that:*

1. $\mu \ge \mathbf{0}$

2. $\nabla E(\mathbf{f}^\star) + \sum_{i=1}^{K} \lambda_i \nabla h_i(\mathbf{f}^\star) + \sum_{j=1}^{L} \mu_j \nabla p_j(\mathbf{f}^\star) = \mathbf{0}$

3. $\mu^T \mathbf{p}(\mathbf{f}^\star) = 0$

3.8.2 Problem Solution

The Theorem of Karush-Kuhn-Tucker can be used to solve this problem. Note the following equations:

$$h_i(\mathbf{f}^\star) = \|\mathbf{K}_i\mathbf{Df}^\star\|^2 - \sigma_i^2 = \mathbf{f}^{\star T}\mathbf{D}^T\mathbf{K}_i^T\mathbf{K}_i\mathbf{Df}^\star - \sigma_i^2$$

$$= \mathbf{f}^{\star T}\mathbf{D}^T\mathbf{K}_i\mathbf{Df}^\star - \sigma_i^2 \tag{3.34}$$

$$\nabla h_i(\mathbf{f}^\star) = 2\mathbf{D}^T\mathbf{K}_i\mathbf{Df}^\star \tag{3.35}$$

$$p_j(\mathbf{f}^\star) = -f_j^\star \tag{3.36}$$

$$\nabla p_j(\mathbf{f}^\star) = -\mathbf{e}_j \tag{3.37}$$

where

$$\mathbf{e}_j \in \Re^L; [\mathbf{e}_j]_i = \begin{cases} 1, & i = j \\ 0, & i \neq j \end{cases} \tag{3.38}$$

Note that $\nabla h_i(\mathbf{f}^\star), \nabla p_j(\mathbf{f}^\star) \in \Re^L$ for each $1 \leq i \leq K, j \in J\{\mathbf{f}^\star\}$.

If we assume $\mathbf{f}^\star$ is a regular point, the solution to the problem can be obtained as follows:

$$E(\mathbf{f}^\star) = \|\mathbf{g} - \mathbf{H}\mathbf{f}^\star\|^2 = (\mathbf{g} - \mathbf{H}\mathbf{f}^\star)^T(\mathbf{g} - \mathbf{H}\mathbf{f}^\star)$$

$$= \mathbf{g}^T\mathbf{g} - 2\mathbf{g}^T\mathbf{H}\mathbf{f}^\star + \mathbf{f}^{\star T}\mathbf{H}^T\mathbf{H}\mathbf{f}^\star \tag{3.39}$$

Note that we have used the fact that $\mathbf{f}^{\star T}\mathbf{H}\mathbf{g} = (\mathbf{g}^T\mathbf{H}\mathbf{f}^\star)^T = \mathbf{g}^T\mathbf{H}\mathbf{f}^\star$ since this quantity is a scalar.

$$\nabla E(\mathbf{f}^\star) = 2\mathbf{H}^T\mathbf{H}\mathbf{f}^\star - 2\mathbf{H}^T\mathbf{g} \tag{3.40}$$

By KKT Condition 2 we have:

$$\nabla E(\mathbf{f}^\star) + \sum_{i=1}^{K} \lambda_i \nabla h_i(\mathbf{f}^\star) + \sum_{j=1}^{L} \mu_j \nabla g_j(\mathbf{f}^\star) = 0$$

$$2\mathbf{H}^T\mathbf{H}\mathbf{f}^\star - 2\mathbf{H}^T\mathbf{g} + 2\sum_{i=1}^{K} \lambda_i \mathbf{D}^T\mathbf{K}_i\mathbf{D}\mathbf{f}^\star - \sum_{j=1}^{L} \mu_i\mathbf{e}_j = 0$$

$$2\left(\mathbf{H}^T\mathbf{H} + \sum_{i=1}^{K} \lambda_i\mathbf{D}^T\mathbf{K}_i\mathbf{D}\right)\mathbf{f}^\star = 2\mathbf{H}^T\mathbf{g} - \sum_{j=1}^{L} \mu_i\mathbf{e}_j$$

Therefore,

$$\mathbf{f}^\star = \left[\mathbf{H}^T\mathbf{H} + \sum_{i=1}^{K} \lambda_i\mathbf{D}^T\mathbf{K}_i\mathbf{D}\right]^{-1}\left(\mathbf{H}^T\mathbf{g} - \frac{1}{2}\sum_{j=1}^{L} \mu_i\mathbf{e}_j\right) \tag{3.41}$$

or alternately, $\mathbf{f}^\star = \left[\mathbf{H}^T\mathbf{H} + \sum_{i=1}^{K} \lambda_i\mathbf{D}^T\mathbf{K}_i\mathbf{D}\right]^{-1}\left(\mathbf{H}^T\mathbf{g} - \frac{1}{2}\sum_{j \in J\{\mathbf{f}^\star\}} \mu_i\mathbf{e}_j\right)$.

KKT Condition 3 states that $\mu^T g(\mathbf{f}^\star) = -\sum_{j=1}^{L} \mu_i f_j^\star = 0$. If $f_j^\star \neq 0$ then $j \notin J\{\mathbf{f}^\star\}$ and $\mu_i = 0$. If $f_j^\star = 0$, then μ_i can be any value and satisfy KKT Condition 3. In fact, by examining (3.41) we discover that the non-zero values of μ are the solutions to:

$$\frac{1}{2} \sum_{p \in J(\mathbf{f}^\star)} \left[\left(\mathbf{H}^T\mathbf{H} + \sum_{i=1}^K \lambda_i \mathbf{D}^T\mathbf{K}_i\mathbf{D} \right)^{-1} \right]_{jp} \mu_p \qquad (3.42)$$

$$= \left[\left(\mathbf{H}^T\mathbf{H} + \sum_{i=1}^K \lambda_i \mathbf{D}^T\mathbf{K}_i\mathbf{D} \right)^{-1} \mathbf{H}^T\mathbf{g} \right]_j$$

$$\forall j; j \in J(\mathbf{f}^\star)$$

The complexity of this equation should not be surprising. A single pixel constrained to be positive by an active constraint will change the solution point of the problem and affects all the other pixels' values. In this way, it effects the degree to which other pixels need to be constrained in order to be positive. Hence the values of all the elements in μ are closely interrelated.

The equality constraint $\mathbf{h}(\mathbf{f}^\star) = \mathbf{0}$ gives us the extra K equations required to solve for the values of λ. Substituting (3.41) into (3.34) we obtain λ as the solution to:

$$\mathbf{X}^T\mathbf{D}^T\mathbf{K}_i\mathbf{D}\mathbf{X} = \sigma_i^2$$

where

$$\mathbf{X} = \left[\mathbf{H}^T\mathbf{H} + \sum_{i=1}^K \lambda_i \mathbf{D}^T\mathbf{K}_i\mathbf{D} \right]^{-1} \left(\mathbf{H}^T\mathbf{g} - \frac{1}{2} \sum_{j \in J\{\mathbf{f}^\star\}}^L \mu_i\mathbf{e}_j \right) \quad \text{and } 1 \le i \le K$$

$$(3.43)$$

In theory, (3.42) and (3.43) form a set of $K + |J(\mathbf{f}^\star)|$ simultaneous equations that may be solved to determine the elements of λ and μ (where $|J(\mathbf{f}^\star)|$ is the number of active inequality constraints). Once the values of λ and μ are found, the solution to the problem can be found using (3.41). It can be seen that the values of the elements in λ and μ are closely related. This is not surprising since constraints on the high frequency components in each region can reasonably be expected to influence with degree to which pixels need to be constrained to positive. In turn, pixels constrained to be positive will influence the high frequency components of the regions, and hence the elements in λ. In practice the problem is not so complex. The value of λ_i for each region is set mainly by the constant σ_i assigned to the region. If the σ_i for each region is not known beforehand (as is often the case), then values of λ_i may be set in a trial-and-error fashion without using (3.43). Once the optimal values of the λ_i are found, an iterative algorithm which operates on one pixel at a time and constrains each pixel to be positive during each iteration will implicitly calculate the μ_i.

3.8.3   Conditions for KKT Theory to Hold

The last subsection supplied the solution to this problem using the Karush-Kuhn-Tucker Theory. This assumes that the pre-conditions of the theory are satisfied. To be specific, for the discussion in the last section to hold, $\mathbf{f}^\star$ must be a regular point. In this section, we will examine the conditions that need to be imposed on the problem for this to be the case. But first, we require some definitions:

Definition 3.3: Define a distance measure $dis2(i, j), 1 \leq i, j \leq NM$, where $L = NM$, the two-dimensional size of the image. Take two pixels with coordinates $x(a, b)$ and $y(á, b́)$. The lexicographic coordinates of these pixels are given by:

$$i = (a - 1)M + b \text{ and } j = (á - 1)M + b́$$

Then we define $dis2(i, j) = \sqrt{(a - á)^2 + (b - b́)^2}$.

Definition 3.3 merely gives the two-dimensional distance between two pixels specified in lexicographic coordinates.

Next consider the high-pass operator, $\mathbf{D}$. Assuming this operator has an impulse response with a finite region of support, which is much less than $\min(N, M)$, then there exists a constant, Z, such that:

$$[\mathbf{D}]_{ij} = 0, \text{ if } dis2(i, j) > Z, \forall i, j; 1 \leq i, j \leq L$$

This inspires another definition:

Definition 3.4: Consider a pixel, f_j, belonging to region i. That is $j \in \{i\}$. We say that f_j is an interior point of region i in terms of $\mathbf{D}$ if and only if $\forall f_p; p \notin \{i\}, dis2(j, p) > Z$, where Z is the smallest number such that $\forall i, j; 1 \leq i, j \leq L, dis2(i, j) > Z \Rightarrow [\mathbf{D}]_{ij} = 0$.

Definition 3.4 states that an interior point of a region in terms of a particular high-pass operator has no neighbors within the region of support of the operator that belong to a different region.

Using these two definitions, we have the following theory:

Theorem 3.3: $\mathbf{f}^\star$ is a regular point if each region $1 \leq i \leq K$ has at least one interior point, r, in terms of $\mathbf{D}$ such that:
$$[\nabla h_i(\mathbf{f}^\star)]_r \neq 0 \text{ and } f_r^\star \neq 0$$
Note that $f_r^\star \neq 0$ is the same as $r \notin J(\mathbf{f}^\star)$.

Proof:

Theorem 3.3 will be proven if we show that a problem satisfying the conditions of the theorem results in $\nabla h_i(\mathbf{f}^\star), \nabla p_j(\mathbf{f}^\star), 1 \leq i \leq K, j \in J(\mathbf{f}^\star)$ being a set of linearly independent vectors.

The theorem can be proven in three parts:

1. Prove that the set $\nabla p_j(\mathbf{f}^\star), j \in J(\mathbf{f}^\star)$ is a set of linearly independent vectors.

2. Prove that the set $\nabla h_i(\mathbf{f}^\star), 1 \leq i \leq K$ is a set of linearly independent vectors.

3. Prove that each $\nabla p_j(\mathbf{f}^\star)$ is linearly independent of the set $\nabla h_i(\mathbf{f}^\star), 1 \leq i \leq K$.

Part (1)

Proof of part (1) is trivial since $\nabla p_j(\mathbf{f}^\star) = -\mathbf{e}_j$ where $-\mathbf{e}_j$ is given by (3.38).

Part (2)

Assume that r is an interior point of region i in terms of $\mathbf{D}$ and $[\nabla h_i(\mathbf{f}^\star)]_r \neq 0$. Consider the rth component of $[\nabla h_j(\mathbf{f}^\star)]_r$ where $j \neq i$.

$$[\nabla h_j(\mathbf{f}^\star)]_r = 2\mathbf{D}^T\mathbf{K}_j\mathbf{Df}^\star = 2\mathbf{D}^T\mathbf{A}$$

where $\mathbf{A} = \mathbf{K}_j\mathbf{Df}^\star$.

$[\nabla h_j(\mathbf{f}^\star)]_r = \sum_{l=1}^{L}[\mathbf{D}]_{rl}\mathbf{A}_l$ since $\mathbf{D}^T = \mathbf{D}$.

If $l \notin \{j\}$ then $[\mathbf{A}]_l = 0$ by the definition of $\mathbf{K}_i$ (3.33).

Therefore $[\nabla h_j(\mathbf{f}^\star)]_r = \sum_{l\in\{j\}}[\mathbf{D}]_{rl}\mathbf{A}_l$.

However, $\text{dis2}(r,l) > Z$ when $l \in \{j\}$ by the definition of an interior point in terms of $\mathbf{D}$; therefore $[\nabla h_j(\mathbf{f}^\star)]_r = 0$.

Since $[\nabla h_i(\mathbf{f}^\star)]_r \neq 0$ and $[\nabla h_j(\mathbf{f}^\star)]_r = 0, \forall j; j \neq i$ we conclude that $\nabla h_i(\mathbf{f}^\star)$ is linearly independent of the other vectors in the set, since it will always have at least one non-zero component for which the other vectors in the set will have zero. If each region in the problem satisfies the conditions of the theorem, the set $\nabla h_i(\mathbf{f}^\star), 1 \leq i \leq K$ will be a set of linearly independent vectors.

Part (3)

Consider a point $r \in \{i\}$ such that $[\nabla h_i(\mathbf{f}^\star)]_r \neq 0$ and $r \notin J(\mathbf{f}^\star)$.

Since r is not a part of the active inequality constraint set, then

$$\nabla h_i(\mathbf{f}^\star) \neq \sum_{j\in J(\mathbf{f}^\star)} \alpha_j\mathbf{e}_j$$

Since $\nabla p_j(\mathbf{f}^\star) = -\mathbf{e}_j$, therefore $\nabla h_i(\mathbf{f}^\star) \neq \sum_{j\in J(\mathbf{f}^\star)} \acute{\alpha}_j\nabla p_j(\mathbf{f}^\star)$.

If every region i has a point that satisfies the conditions of Theorem 3.3, then the set $\nabla p_j(\mathbf{f}^\star), j \in J(\mathbf{f}^\star)$ is linearly independent to the set $\nabla h_i(\mathbf{f}^\star), 1 \leq i \leq K$.

Hence the proof is completed. We have shown that the set $\{\nabla h_i(\mathbf{f}^\star), \nabla p_j(\mathbf{f}^\star)\}$, $1 \leq i \leq K, j \in J(\mathbf{f}^\star)$ is a set of linearly independent vectors.

This section has shown that if an image can be divided into a number of regions each satisfying Theorem 3.3, the Theorem of Karush-Kuhn-Tucker states that solutions to this problem will exist.

3.8.4   Discussion

Various schemes such as measuring local statistical properties may be used to select the regions and assign values of λ to these regions. To use the analysis in this section, we then need to make sure that the selected regions satisfy the conditions of Theorem 3.3. Note that the conditions imposed on the regions by Theorem 3.3 are *sufficient* but not *necessary*. This means that regions satisfying these conditions will be able to be analyzed by the approach in this section, but regions that do not satisfy these conditions may also be just as valid. Note that the algorithms derived by the penalty function approach in Section 3.3 are not compatible with the analysis in this section. The analysis in this section expands the single constraint problem into the problem of assigning a different constraint value to each of a number of predetermined regions and Theorem 3.3 is not valid if the regions are too small or too narrow. However, the analysis in Section 3.3 began by assigning each pixel a different value of λ and from this beginning moved into the concept of a region. The concept of regions in Section 3.3 was shown to arise naturally, but is not as integral a concept as it was in this section.

We will not delve any more into the Karush-Kuhn-Tucker analysis since although it is more mathematically justified than the penalty function concept, we have shown in this chapter that algorithms based on the penalty function concept produce acceptable results and have no conditions upon the size and shape of the regions created.

3.9   Summary

In this chapter the use of spatially adaptive weights in constrained restoration methods was considered.

The first use of spatially adaptive weights we considered was adaptation of the neural weights to restore images degraded by spatially variant distortions. We showed how the neural network approach could be used to solve this problem without a great deal of extra computations.

The second use of spatially adaptive weights we examined was in order to implement a spatially adaptive constraint parameter. Since the human visual system favors the presence of edges and boundaries, rather than more subtle differences in intensity in homogeneous areas [85], noise artifacts may be less disturbing in high contrast regions than in low contrast regions. It is then advantageous to use a stronger constraint in smooth areas of an image than in high-contrast regions. While traditional restoration methods find it difficult to implement an adaptive restoration spatially across an image, neural network-based image restoration methods are particularly amenable to spatial variance of the restoration parameters.

We expanded upon the basic restoration functional by looking at it as having a restoration term followed by a penalty term. From this analysis, two methods of selecting the best constraint parameter for each pixel were investigated. The best method was based on using local image statistics to select the optimal value of regularization parameter. This method imitates the human visual system and produces superior results when compared to non-adaptive methods. In addition it was found that no disadvantage occurred when the values of regularization parameter for each pixel were chosen by the restoration algorithm before starting the restoration, rather than during each iteration of the restoration procedure. In fact, pre-computing of the regularization parameter further increased the restoration speed.

In the next section of this chapter, the work was expanded upon to adaptively restore images degraded by a spatially variant PSF. It was shown how adaptive regularization techniques can compensate for insufficient knowledge of the degradation in the case of spatially variant distortions. Moreover, an adaptive spatially variant restoration is shown to be able to be completed in the same order of magnitude of time as a much simpler non-adaptive spatially invariant restoration.

In the course of this chapter, we have examined some implementation considerations and tested the algorithms developed on real imagery.

We finished our analysis of the adaptive constraint concept by looking at how the basic restoration functional may be expanded upon using the theory of Lagrange. We looked at what conditions the theorem of Lagrange applied to the adaptive constraint problem and how these conditions related back to the penalty function-based approaches.

Chapter 4

Perceptually Motivated Image Restoration

4.1   Introduction

In Chapter 3, we mentioned the problems caused by considering the image to be an ensemble of stationary processes. Any restoration process based on this concept can only ever produce sub-optimal results. However, there is another consideration. When a restoration algorithm is designed with the aim of creating an image which will be more pleasing to the human eye, we must incorporate some kind of model of the human visual system. At the basis of most restoration algorithms is some form of image error measure which is being minimized.

The most common method to compare the similarity of two images is to compute their mean square error (MSE). However, the MSE relates to the power of the error signal and has little relationship to human visual perception. An important drawback to the MSE and any cost function which attempts to use the MSE to restore a degraded image is that the MSE treats the image as a stationary process. All pixels are given equal priority regardless of their relevance to human perception. This suggests that information is ignored. When restoring images for the purpose of better clarity as perceived by humans the problem becomes acute. Considerations regarding human perception have been examined in the past [82, 83, 85, 86, 87, 88, 89, 90, 91, 92, 93]. When humans observe the differences between two images, they do not give much consideration to the differences in individual pixel level values. Instead humans are concerned with matching edges, regions and textures between the two images. This is contrary to the concepts involved in the MSE. From this it can be seen that any cost function which treats an image as a stationary process can only produce a sub-optimal result.

In Chapter 1, we looked at a new error measure based on comparing local variances which examines the image in a regional sense rather than a pixel-by-pixel sense. This error measure uses some simple concepts behind human perception without being too complex. In this chapter, we further develop these concepts by examining two restoration algorithms which incorporate versions of this image error measure. The cost functions that these algorithms are based on are non-linear and cannot be efficiently implemented by conventional methods. In this chapter we introduce extended neural network algorithms to iteratively perform the restoration. We show that the new cost functions and processing algorithms perform very well when applied to both color and grayscale images. One important property of the methods we will examine in this chapter, compared with the neural network implementation of the constrained least square filter, is that they are very fault-tolerant in the sense that when some of the neural connections are damaged, these algorithms can still produce very satisfactory results. Comparison with some of the conventional methods will be provided to justify the new methods.

This chapter is organized as follows. Section 4.2 describes the motivation for incorporating the error measure described in Chapter 1 into a cost function. Section 4.3 presents the restoration cost function incorporating a version of the proposed image error measure from Chapter 1. Section 4.4 builds on the previous section to present an algorithm based on a more robust variant of the novel image error measure. Section 4.5 describes implementation considerations. Section 4.6 presents some numerical data to compare the algorithms presented in this chapter with those of previous chapters, and Section 4.7 summarizes this chapter.

4.2   Motivation

In the introduction to this chapter, we considered the problems inherent in using the MSE (Mean Square Error) and SNR (Signal to Noise Ratio) to compare two images. It was seen that the MSE and SNR have little relationship to the way that humans perceive the differences between two images. Although incorporating concepts involved in human perception may seem a difficult task, a new image error measure was presented in Chapter 1 which, despite its simplicity, incorporates some concepts involved in human appraisal of images.

In Chapter 3, the basic neural network restoration algorithm described in Chapter 2 was expanded to restore images adaptively using simple human visual concepts. These adaptive algorithms obtained superior results when compared to the non-adaptive algorithm, and were shown to produce a more robust restoration when errors occurred due to insufficient knowledge of the degrading function. Despite the improved performance of the adaptive algorithms, it is still not simple to choose the correct values of the constraint parameter, λ. In the case of the adaptive algorithms described in Chapter 3, the problem is compounded by the fact that many values of λ must be selected, rather than just one value as in the non-adaptive case. In addition, the selected λ values must be related to local variance levels.

We desire to create an algorithm which can adaptively restore an image, using simple concepts involved in human perception, with only a few free parameters to be set. Such an algorithm would be more robust and easier to use than the variance selection algorithm in Chapter 3. In the previous adaptive algorithm, minimizing the MSE was still at the base of the restoration strategy. However, Chapter 1 provides us with a simple alternative to the MSE. It seems logical to create a new cost function which minimizes a LSMSE related term. In this way, the adaptive nature of the algorithm would be incorporated in the LSMSE term rather than imposed by the external selection of λ values.

4.3   A LVMSE-Based Cost Function

The discussion in the previous section prompts us to structure a new cost function which can properly incorporate the LSMSE into restoration. Since the formula for the neural network cost function has a term which attempts to minimize the MSE, let us look at restoring images using a cost function with a term which endeavors to minimize a LSMSE-related error measure. Let's start by adding an additional term to (2.3). The new term evaluates the local variance mean square error [118, 102]:

$$\text{LVMSE} = \frac{1}{NM} \sum_{x=0}^{N-1} \sum_{y=0}^{M-1} (\sigma_A^2(\mathbf{f}(x,y)) - \sigma_A^2(\mathbf{g}(x,y)))^2 \qquad (4.1)$$

where $\sigma_A^2(\mathbf{f}(x,y))$ is the variance of the local region surrounding pixel (x,y) in the first image and $\sigma_A^2(\mathbf{g}(x,y))$ is the variance of the local region surrounding pixel (x,y) in the image with which we wish to compare the first image.

Hence the new cost function we suggest is:

$$E = \frac{1}{2}\|\mathbf{g} - \mathbf{H}\hat{\mathbf{f}}\|^2 + \frac{\lambda}{2}\|\mathbf{D}\hat{\mathbf{f}}\|^2 + \frac{\theta}{NM} \sum_{x=0}^{N-1} \sum_{y=0}^{M-1} (\sigma_A^2(\hat{\mathbf{f}}(x,y)) - \sigma_A^2(\mathbf{g}(x,y))^\star)^2 \quad (4.2)$$

In this case, $\sigma_A^2(\hat{\mathbf{f}}(x,y))$ is the variance of the local region surrounding pixel (x,y) in the image estimate and $\sigma_A^2(\mathbf{g}(x,y))^\star$ is the variance of the local region surrounding pixel (x,y) in the degraded image scaled to predict the variance in the original image. The comparison of local variances rather than local standard deviations was chosen for the cost function since it is easier and more efficient to calculate. The first two terms in (4.2) ensure a globally balanced restoration, whereas the added LVMSE term enhances local features. In (4.2), $\sigma_A^2(\mathbf{g}(x,y))^\star$ is determined as follows. Since the degraded image has been blurred, image variances in $\mathbf{g}$ will be lower than the corresponding variances in the original image. In this case, the variances $\sigma_A^2(\mathbf{g}(x,y))^\star$ would be scaled larger than $\sigma_A^2(\mathbf{g}(x,y))$ to reflect the decrease in variance due to the blurring function. In general, if we consider an image degraded by a process which is modeled by (2.2), then we find that a useful approximation is:

$$\sigma_A^2(\mathbf{g}(x,y))^\star = K(x,y)(\sigma_A^2(\mathbf{g}(x,y)) - J(x,y)) \qquad (4.3)$$

where $J(x, y)$ is a function of the noise added to the degraded image at point (x, y) and $K(x, y)$ is a function of the degrading point spread function at point (x, y). Although it may appear difficult to accurately determine the optimal values of $K(x, y)$, in fact, as we will demonstrate later, the algorithm is extremely tolerant of variations in this factor and only a rough estimate is required. For example, if the image degradation is a moderate blurring function, with a region of support of around 5 or 7, then $K(x, y)$ would be set to 2 for all pixels in the image. This indicates that the local variances in the original image are on average approximately twice that of the degraded image. A high degree of accuracy is not required. In highly textured regions of the image where the preservation of image details are most important, the LVMSE term requires that the variance of the region be large, and the first two terms of (4.2) ensure the sharpness and accuracy of the image features.

4.3.1   The Extended Algorithm for the LVMSE-Modified Cost Function

The LVMSE modified cost function does not fit easily into the neural network energy function as given by (2.4); however, an efficient algorithm can be designed to minimize this cost function. One of the first considerations when attempting to implement the LVMSE cost function is prompted by a fundamental difference in the cost function which occurs due to the addition of the new term. In the case of a cost function based on minimizing mean square error alone, any changes in an individual pixel's value affects the entire image MSE in a simple way. The square error between any pixel in the image estimate and the corresponding pixel in the original image does not affect the square error of its neighbors. This simplifies the implementation of the cost function. In the case of the LVMSE modified cost function, it is different. When a pixel's value is altered by the algorithm, the total change in the LVMSE is not a simple function of the current pixel's change in value alone. Changing the current pixel's value changes its own local variance, and the local variances of all of its neighbors within an A by A proximity of the current pixel. Hence to truly calculate the total change in LVMSE for the entire image, the algorithm must calculate how changing the current pixel's value affects the local variances of all its neighbors and how these changes effect the overall LVMSE. This approach is computationally prohibitive. To resolve this problem we must go back to the fundamental justification for adding the LVMSE term in the first place. The justification for adding this term was the fact that we wished to create a cost function which matched the local statistics of pixels in the original image to that of the image estimate. In this case it is sufficient that the algorithm considers only minimizing the difference in the local variance of the estimated image pixel to the corresponding original image pixel and not minimizing the total LVMSE of the image. The great benefit arising from this approximation will become apparent as explained below.

The first step in the development of the algorithm is a change in notation. For an N by M image let **f** represent the lexicographically organized image vector of length NM as per the model given by (2.2) and the algorithm for the unmodified neural network cost function (2.3). The translation between the two indices x and y of $f(x,y)$ and the single index i of f_i is given by:

$$i = x + yN \qquad (4.4)$$

Define x^k and y^k as the two-dimensional x and y values associated with pixel k by (4.4).

Define the two-dimensional distance between pixels i and j as:

$$\text{dis2́}(i,j) = |x^i - x^j| + |y^i - y^j| \qquad (4.5)$$

Note that $\text{dis2́}(i,j)$ has a lot in common with $\text{dis2}(i,j)$ in Definition 3.2. However, $\text{dis2́}(i,j)$ describes the *city block* distance between two pixels, while $\text{dis2}(i,j)$ describes the *Euclidean* distance between the same two pixels.

Let $\mathbf{\Phi}^i$ represent the NM by NM matrix which has the following property:

$$\text{Let } \mathbf{f}^i = \mathbf{\Phi}^i \mathbf{f} \qquad (4.6)$$

then

$$[\mathbf{f}^i]_j = \begin{cases} 0, & \text{dis2́}(i,j) > \dfrac{A-1}{2} \\ f_j, & \text{dis2́}(i,j) \le \dfrac{A-1}{2} \end{cases} \qquad (4.7)$$

$\mathbf{\Phi}^i$ has the effect of setting to zero all pixels not within the A by A neighborhood centered on the pixel with coordinates x^i, y^i. As a shortened notation we will denote $[\mathbf{f}^i]_j$ as f^i_j. Using this notation, the average pixel value in the A by A region surrounding pixel i is given by:

$$M_A(i) = \frac{1}{A^2} \sum_{j=1}^{MN} f^i_j$$

Let $\beta_i = \sum_{j=1}^{NM} (f^i_j)^2$ and $\gamma_i = \sum_{j=1}^{NM} f^i_j$. Then the estimated variance of the A by A region surrounding pixel i is given by:

$$V^i = \frac{\beta_i}{A^2} - \frac{\gamma_i^2}{A^4} \qquad (4.8)$$

Note that, strictly speaking, V^i is an estimate of the variance of this region given the available pixel values. The true variance of this region is the expectation of the second moment. However, (4.8) is a suitable approximation given the available data. In the rest of this analysis, (4.8) will be called the *local variance* and the term *estimated local variance* will be used to refer to $\sigma_A^2(\mathbf{g}(x,y))^\star$.

The LVMSE between the image estimate, $\hat{\mathbf{f}}$, and the original image, $\mathbf{f}$, may then be written as:

$$\text{LVMSE}(\hat{\mathbf{f}}, \mathbf{f}) = \frac{1}{NM} \sum_{i=1}^{NM} (V^i(\hat{\mathbf{f}}) - V^i(\mathbf{f}))^2 \qquad (4.9)$$

Let $V^i(\mathbf{f})$ be approximated by Vf^i. Vf^i is the estimate of the local variance of pixel i in the original image based on the degraded image and knowledge of the degrading point spread function as per equation (4.3). Vf^i is calculated before the algorithm commences and remains a constant throughout the restoration procedure.

The algorithm we examine in this section first computes the negative direction of the gradient which gives an indication of whether increasing or decreasing the current neuron's value will result in a net decrease in energy. Once the negative gradient is found the neuron's value is changed in unit steps and the resultant energy decrease after each step is computed. This ends when no further energy minimization is possible. In Chapter 2, we showed that the negative gradient of the unmodified cost function, (2.3), is in fact the input to the neuron. Hence the negative gradient of the modified cost function will therefore be the input to the neuron minus the derivative of (4.9).

The gradient of (4.9) is given by:

$$\frac{\delta}{\delta \hat{f_i}}\text{LVMSE} = \frac{2}{NM}(V^i(\hat{\mathbf{f}}) - Vf^i)\frac{\delta}{\delta \hat{f_i}}(V^i(\hat{\mathbf{f}}) - Vf^i) \tag{4.10}$$

Note that this formula is an approximation of the gradient which ignores the contributions of the local variances of the pixels adjacent to i to the overall LVMSE of the image.

$$\frac{\delta V^i(\hat{\mathbf{f}})}{\delta \hat{f_i}} = \frac{2\hat{f_i}}{A^2} - \frac{2\gamma_i}{A^4} \tag{4.11}$$

Note that $\hat{f}^i_i = \hat{f_i}$. Substituting (4.11) into (4.10) we obtain:

$$\frac{\delta}{\delta \hat{f_i}}\text{LVMSE} = \frac{2}{NM}\left(\frac{\beta_i}{A^2} - \frac{\gamma_i^2}{A^4} - Vf^i\right)\left(\frac{2\hat{f_i}}{A^2} - \frac{2\gamma_i}{A^4}\right)$$

therefore

$$\frac{\delta}{\delta \hat{f_i}}\text{LVMSE} = \frac{4}{NMA^2}\left(\frac{\hat{f_i}\beta_i}{A^2} - \frac{\hat{f_i}\gamma_i^2}{A^4} - \hat{f_i}Vf^i - \frac{\beta_i\gamma_i}{A^4} + \frac{\gamma^3}{A^6} + \frac{\gamma_i Vf^i}{A^2}\right) \tag{4.12}$$

Multiplying (4.12) by θ and subtracting it from the input to the neuron gives us the negative gradient of the cost function. Given a change in the value of pixel i, the resultant change in energy is the previous change in energy given by (2.12) plus θ times the change in LVMSE. The change in LVMSE is given by:

$$\Delta\text{LVMSE} = \frac{1}{NM}\left((V^i_{new} - Vf^i)^2 - (V^i_{old} - Vf^i)^2\right) \tag{4.13}$$

where

$$\gamma_i^{new} = \gamma_i^{old} + \Delta\hat{f}_i$$
$$\beta_i^{new} = \beta_i^{old} + 2\hat{f}_i\Delta\hat{f}_i + (\Delta\hat{f}_i)^2$$
$$V_{new}^i = \frac{\beta_i^{new}}{A^2} - \frac{(\gamma_i^{new})^2}{A^4}$$
$$= V_{old}^i + \frac{\left(\Delta\hat{f}_i\right)^2}{A^2} + \frac{2\hat{f}_i\Delta\hat{f}_i}{A^2} - \frac{2\gamma_i^{old}\Delta\hat{f}_i}{A^4} - \frac{\left(\Delta\hat{f}_i\right)^2}{A^4} \qquad (4.14)$$

where β_i^{old}, γ_i^{old} and V_{old}^i are the values of these parameters before the change in the state of neuron i occurred. The LVMSE algorithm is therefore:

Algorithm 4.1.

repeat
{
    for $i = 1, \ldots, L$ do
    {

$$u_i = b_i + \sum_{j=1}^{L} w_{ij}\hat{f}_j$$
$$\beta_i^{old} = \sum_{j=1}^{L} \left(\hat{f}_j^i\right)^2$$
$$\gamma_i^{old} = \sum_{j=1}^{L} \hat{f}_j^i$$
$$V_{old}^i = \frac{\beta_i^{old}}{A^2} - \frac{\left(\gamma_i^{old}\right)^2}{A^4}$$
$$-\frac{\delta E}{\delta \hat{f}_i} = u_i$$
$$-\frac{4\theta}{NMA^2}\left(\frac{\hat{f}_i\beta_i^{old}}{A^2} - \frac{\hat{f}_i\left(\gamma_i^{old}\right)^2}{A^4} - \hat{f}_iVf^i - \frac{\beta_i^{old}\gamma_i^{old}}{A^4} + \frac{\left(\gamma_i^{old}\right)^3}{A^6} + \frac{\gamma_i^{old}Vf^i}{A^2}\right)$$

$$\Delta\hat{f}_i = G\left(-\frac{\delta E}{\delta \hat{f}_i}\right) \text{ where } G(u) = \begin{cases} 1, & u > 0 \\ 0, & u = 0 \\ -1, & u < 0 \end{cases}$$

$$V_{new}^i = V_{old}^i + \frac{\left(\Delta\hat{f}_i\right)^2}{A^2} + \frac{2\hat{f}_i\Delta\hat{f}_i}{A^2} - \frac{2\gamma_i^{old}\Delta\hat{f}_i}{A^4} - \frac{\left(\Delta\hat{f}_i\right)^2}{A^4}$$
$$\Delta E = -\frac{1}{2}w_{ii}\left(\Delta\hat{f}_i\right)^2 - u_i\Delta\hat{f}_i + \frac{\theta}{NM}\left(\left[V_{new}^i - Vf^i\right]^2 - \left[V_{old}^i - Vf^i\right]^2\right)$$
repeat
{
$$\hat{f}_i(t+1) = K(\hat{f}_i(t) + \Delta\hat{f}_i)$$
$$\text{where } K(u) = \begin{cases} 0, & u < 0 \\ u, & 0 \leq u \leq S \\ S, & u > S \end{cases}$$
$$u_i = u_i + w_{ii}\Delta\hat{f}_i$$

$$V_{old}^i = V_{new}^i$$
$$\gamma_i^{old} = \gamma_i^{old} + \Delta \hat{f}_i$$

$$V_{new}^i = V_{old}^i + \frac{(\Delta \hat{f}_i)^2}{A^2} + \frac{2\hat{f}_i \Delta \hat{f}_i}{A^2} - \frac{2\gamma_i^{old} \Delta \hat{f}_i}{A^4} - \frac{(\Delta \hat{f}_i)^2}{A^4}$$

$$\Delta E = -\frac{1}{2} w_{ii}(\Delta \hat{f}_i)^2 - u_i \Delta \hat{f}_i + \frac{\theta}{NM}([V_{new}^i - Vf^i]^2 - [V_{old}^i - Vf^i]^2)$$

until $\Delta E \geq 0$

$\qquad$ }

$\qquad$ $t = t + 1$

}

until $\left(\hat{f}_i(t) = \hat{f}_i(t-1) \forall i = 1, \ldots, L \right)$

Note that Algorithm 4.1 still utilizes some features of Algorithm 2.3, specifically the use of bias inputs and interconnection strength matrices.

4.3.2   Analysis

It is important to verify that Algorithm 4.1 acts upon a pixel in the intended manner. To verify this, we must examine the LVMSE term in Algorithm 4.1 more closely.

According to (4.12), the gradient of the LVMSE term of the cost function, when all pixels except $\hat{f}_i$ are held constant, is given by:

$$\frac{\delta}{\delta \hat{f}_i} \text{LVMSE} = \frac{2}{NM} \left(\frac{\beta_i}{A^2} - \frac{\gamma_i^2}{A^4} - Vf^i \right) \left(\frac{2\hat{f}_i}{A^2} - \frac{2\gamma_i}{A^4} \right)$$

Note that $\beta_i = \sum_{j=1}^{L}(f_j^i)^2$ and $\gamma_i = \sum_{j=1}^{L} f_j^i$ can be rewritten as $\beta_i = \acute{\beta}_i + \hat{f}_i^2$ and $\gamma_i = \acute{\gamma}_i + \hat{f}_i$ where $\acute{\beta}_i = \sum_{j=1, j\neq i}^{L}(f_j^i)^2$ and $\acute{\gamma}_i = \sum_{j=1, j\neq i}^{L} f_j^i$. In this way, we can extract the elements of (4.12) that depend on $\hat{f}_i$. Hence we obtain:

$$\frac{\delta}{\delta \hat{f}_i} \text{LVMSE} = \frac{2}{NM} \left(\frac{\hat{f}_i^2}{A^2} + \frac{\acute{\beta}_i}{A^2} - \frac{(\hat{f}_i + \acute{\gamma}_i)^2}{A^4} - Vf^i \right) \left(\frac{2\hat{f}_i}{A^2} - \frac{2\hat{f}_i}{A^4} - \frac{2\acute{\gamma}_i}{A^4} \right)$$

$$= \frac{2}{NM} \left(\frac{\hat{f}_i^2}{A^2} - \frac{\hat{f}_i^2}{A^4} - \frac{2\hat{f}_i \acute{\gamma}_i}{A^4} + \frac{\acute{\beta}_i}{A^2} - \frac{\acute{\gamma}_i^2}{A^4} - Vf^i \right) \left(\frac{2\hat{f}_i}{A^2} - \frac{2\hat{f}_i}{A^4} - \frac{2\acute{\gamma}_i}{A^4} \right) \quad (4.15)$$

Observe that $\frac{\acute{\beta}_i}{A^2} - \frac{\acute{\gamma}_i^2}{A^4}$ is an approximation to the local variance at pixel i neglecting the contribution of the value of pixel i itself. Let $\acute{V} = \frac{\acute{\beta}_i}{A^2} - \frac{\acute{\gamma}_i^2}{A^4}$. As A increases in value, $\acute{V}$ approaches the value of the local variance at pixel i. Similarly, we can define an approximation to the local mean of pixel i as:

$$\acute{M} = \frac{1}{A^2} \sum_{j=1, j\neq i}^{NM} \hat{f}_j^i = \frac{\acute{\gamma}_i}{A^2} \quad (4.16)$$

If A is greater than 3, then pixel i contributes less than 15% of the value of the local mean and local variance of its neighborhood. If A is 7 then the contribution is only 2%. Hence approximating $\acute{V}$ as the local variance is valid.

This leaves us with:

$$\frac{\delta}{\delta \hat{f}_i} \text{LVMSE} = \frac{4}{NMA^2} \left(\hat{f}_i^2 \left[\frac{1}{A^2} - \frac{1}{A^4} \right] - \frac{2 \hat{f}_i \acute{M}}{A^2} + J \right) \left(\hat{f}_i \left[1 - \frac{1}{A^2} \right] - \acute{M} \right)$$

$$(4.17)$$

where $J = \acute{V} - V f^i$.

The points for which (4.17) is equal to zero are the stationary points of the LVMSE term in (4.2). Equation (4.17) is zero when $\hat{f}_i = \acute{M}/(1 - \frac{1}{A^2})$. This corresponds to the case where $\hat{f}_i$ is approximately equal to its local mean. Equation (4.17) also has zeroes when $\hat{f}_i$ satisfies:

$$\hat{f}_i = \frac{\dfrac{2\acute{M}}{A^2} \pm \sqrt{\dfrac{4\acute{M}^2}{A^4} - 4J \left(\dfrac{1}{A^2} - \dfrac{1}{A^4} \right)}}{2 \left(\dfrac{1}{A^2} - \dfrac{1}{A^4} \right)}$$

If A is greater than or equal to 5, then the error resulting from approximating $1/A^2 - 1/A^4$ as $1/A^2$ is less than 4%. Therefore, if we assume that A is large enough that $1/A^2 \gg 1/A^4$ then (4.17) has zeroes at:

$$\hat{f}_i \approx \acute{M} \pm A \sqrt{\frac{\acute{M}^2}{A^2} - J} \qquad (4.18)$$

Note that (4.18) indicates that if $J > 0$, (4.18) may not have a real-valued solution. There are three cases to examine.

CASE 1: $J < 0$

When $J < 0$, then local variance of the region surrounding pixel i is less than the estimated local variance. Equations (4.18) and (4.17) indicate that the LVMSE term in the cost function will have three stationary points. The function will then appear as Figure 4.1. The stationary point given by $\hat{f}_i = \acute{M}/(1 - \frac{1}{A^2})$ becomes a maximum and the function has two minimum given by (4.18). At least one of these minimum will be in the domain $\hat{f}_i > 0$, and so as long as J is not excessively negative it is possible to minimize the LVMSE term in the cost function in this case.

This is as one would expect since if $J < 0$, then the variance of the local region surrounding pixel i needs to be increased to meet the target variance estimate. It is always possible to increase the local variance of pixel i by moving that pixel's value further from the mean. This is why the stationary point $\hat{f}_i = \acute{M}/(1 - \frac{1}{A^2})$, which is the point where $\hat{f}_i$ is approximately equal to the mean of the local region, becomes a maximum.

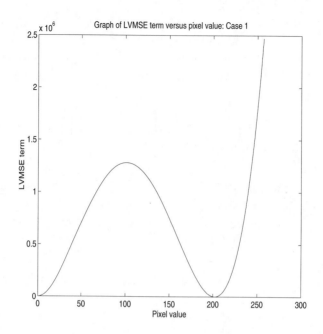

Figure 4.1: Graph of function for Case 1

CASE 2: $0 < J \le \frac{\acute{M}^2}{A^2}$

When $0 < J \le \frac{\acute{M}^2}{A^2}$, then the local variance of region surrounding pixel i is greater than the estimated local variance, but the difference is not great. Equations (4.18) and (4.17) indicate that the LVMSE term in the cost function will again have three stationary points. The function will then appear as Figure 4.2.

The stationary point given by $\hat{f}_i = \acute{M}/(1 - \frac{1}{A^2})$ becomes a maximum and the function has two minimum given by (4.18). At least one of these minima will be in the domain $\hat{f}_i > 0$, and it is again possible to minimize the LVMSE term in the cost function in this case.

When $J > 0$, the local variance of pixel i needs to be decreased to match the target variance estimate. If $J \le \frac{\acute{M}^2}{A^2}$, it is possible to match the local variance of pixel i with the target variance estimate by moving that pixel's value toward the mean pixel value in the local region. The stationary point $\hat{f}_i = \acute{M}/(1 - \frac{1}{A^2})$, which is the point where $\hat{f}_i$ is approximately equal to the mean of the local region, is again a maximum, unless $J = \frac{\acute{M}^2}{A^2}$. When $J = \frac{\acute{M}^2}{A^2}$, all three stationary points correspond to the same value and the function has one minimum at $\hat{f}_i = \acute{M}/(1 - \frac{1}{A^2}) \approx \acute{M}$.

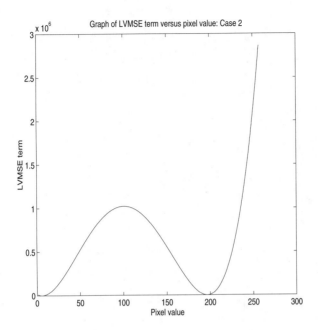

Figure 4.2: Graph of function for Case 2

CASE 3: $J > \frac{\acute{M}^2}{A^2}$

When $J > \frac{\acute{M}^2}{A^2}$, then the local variance of region surrounding pixel i is greater than the estimated local variance, but the difference is too large for equality to be reached by changing a single pixel's value. Equation (4.18) will have no solutions and the LVMSE term in the cost function will have one stationary point. The function will then appear as Figure 4.3.

The stationary point given by $\hat{f}_i = \acute{M}/(1 - \frac{1}{A^2})$ becomes a minimum, where $\acute{M} > 0$ since all pixel values are constrained to be positive. So the minimum will be in the domain $\hat{f}_i > 0$, and it is possible to minimize the LVMSE term in the cost function in this case.

When $J > \frac{\acute{M}^2}{A^2}$, the local variance of pixel i needs to be decreased to match the target variance estimate. The minimum local variance is obtained by decreasing the value of the current pixel to the mean value of the local region. In this case, the minimum local variance obtainable by altering one pixel's value is still greater than the estimate. The stationary point $\hat{f}_i = \acute{M}/(1 - \frac{1}{A^2})$ is the point where $\hat{f}_i$ is approximately equal to the mean of the local region. Thus, the pixel will move toward the mean value of the local region, hence reducing the local variance as much as possible.

By examining each of the three cases above we can see that the LVMSE term in the cost function is well behaved despite its non-linear nature. In all cases a minimum exists and the local variance of the current pixel's region will always move closer to the target variance estimate.

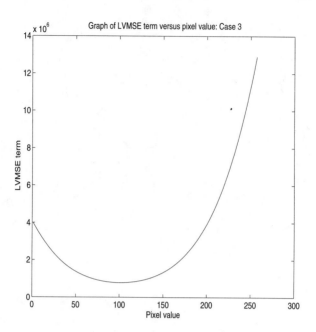

Figure 4.3: Graph of function for Case 3

4.4   A Log LVMSE-Based Cost Function

The previous section showed that the LVMSE cost term in (4.2) is well behaved, and results in an algorithm which does indeed work as intended to match the local variance of the region surrounding the current pixel to that of a target variance estimate of the original image. The LVMSE term in (4.2) has its greatest effect when the difference between the actual local variance and the target variance estimate is large. When the difference is small, the LVMSE term in (4.2) has little effect. The strength of the LVMSE term is proportional to the square of the absolute difference between the two variances and does not depend on the level of the variances. The dependence on the absolute difference between the variances is in fact a disadvantage. When the local variance of a region is large and the target variance for that region is also large, then noise will not be readily noticed. In this case, the first term in the cost function which ensures sharpness of edges should be allowed to dominate the restoration. However, in this case, the difference between the target variance and actual variance may be large, causing the LVMSE term to dominate instead. On the other hand, when the local variance of a region is small and the target variance for that region is also small, we would want the LVMSE term to dominate, since this term would keep the local variance low and suppress noise. However, in this case since the target variance and the actual variance are small, the LVMSE term can also be too small and have an insufficient effect.

This prompts us to move away from absolute differences in local variances, and, instead, compare the ratio of the local variance and its estimate. Taking this idea one step further, we notice that taking the log of this ratio provides additional emphasis of small differences in variance at low variance levels and de-emphasis of large differences in variance at high variance levels.

Hence a new cost function is introduced [102]:

$$E = \frac{1}{2}\|\mathbf{g} - \mathbf{H}\hat{\mathbf{f}}\|^2 + \frac{\lambda}{2}\|\mathbf{D}\hat{\mathbf{f}}\|^2 + \frac{\theta}{NM}\sum_{x=0}^{N-1}\sum_{y=0}^{M-1}\left[\ln\left(\frac{\sigma_A^2(\hat{\mathbf{f}}(x,y))}{\sigma_A^2(\mathbf{g}(x,y))^\star}\right)\right]^2 \quad (4.19)$$

We will denote this new term in (4.19) the Log Local Variance Ratio (Log LVR) term.

4.4.1 The Extended Algorithm for the Log LVR-Modified Cost Function

The algorithm to minimize (4.19) has the same basic strategy as the algorithm created in Section 4.3. First, the negative direction of the gradient is computed. Once the negative gradient is found, the neuron value is changed in unit steps and the resultant energy decrease after each step is computed. This ends when no further energy reduction is possible. As in the last section, multiplying the partial derivative of the Log LVR term in (4.19) by θ and subtracting it from the input to the neuron gives us the negative gradient of the cost function. As defined in Section 4.3, the local variance of the A by A region centered on pixel i is given by:

$$V^i = \frac{\beta_i}{A^2} - \frac{\gamma_i^2}{A^4}$$

where $\beta_i = \sum_{j=1}^{L}(f_j^i)^2$ and $\gamma_i = \sum_{j=1}^{L}f_j^i$.

The gradient of the Log LVR term in (4.19) is given by:

$$\frac{\delta}{\delta\hat{f}_i}\text{ Log LVR} = \frac{1}{NM}\,2\ln\left(\frac{V^i}{Vf^i}\right)\frac{Vf^i}{V^i}\frac{1}{Vf^i}\left(\frac{2\hat{f}_i}{A^2} - \frac{2\gamma_i}{A^4}\right)$$

$$= \frac{4}{NM}\ln\left(\frac{\frac{\beta_i}{A^2} - \frac{\gamma_i^2}{A^4}}{Vf^i}\right)\frac{\left(\hat{f}_i - \frac{\gamma_i}{A^2}\right)}{\left(\beta_i - \frac{\gamma_i^2}{A^2}\right)} \quad (4.20)$$

Similarly, given a change in the value of pixel i, the resultant change in energy is the previous change in energy given by (2.12) plus θ times the change in Log LVR. The change in Log LVR is given by:

$$\Delta \text{ Log LVR} = \frac{1}{NM}\left(\ln\left[\frac{V_{new}^i}{Vf^i}\right]^2 - \ln\left[\frac{V_{old}^i}{Vf^i}\right]^2\right) \quad (4.21)$$

where

$$\gamma_i^{new} = \gamma_i^{old} + \Delta \hat{f}_i$$

$$\beta_i^{new} = \beta_i^{old} + 2\hat{f}_i \Delta \hat{f}_i + (\Delta \hat{f}_i)^2$$

$$V_{new}^i = \frac{\beta_i^{new}}{A^2} - \frac{(\gamma_i^{new})^2}{A^4}$$

$$= V_{old}^i + \frac{\left(\Delta \hat{f}_i\right)^2}{A^2} + \frac{2\hat{f}_i \Delta \hat{f}_i}{A^2} - \frac{2\gamma_i^{old} \Delta \hat{f}_i}{A^4} - \frac{\left(\Delta \hat{f}_i\right)^2}{A^4}$$

The new algorithm is therefore:

Algorithm 4.2.
repeat
  {
  for $i = 1, \ldots, L$ do
  {

$$u_i = b_i + \sum_{j=1}^{L} w_{ij} \hat{f}_j$$

$$\beta_i^{old} = \sum_{j=1}^{L} \left(\hat{f}_j^i \right)^2$$

$$\gamma_i^{old} = \sum_{j=1}^{L} \hat{f}_j^i$$

$$V_{old}^i = \frac{\beta_i^{old}}{A^2} - \frac{(\gamma_i^{old})^2}{A^4}$$

$$-\frac{\delta E}{\delta \hat{f}_i} = u_i - \frac{4\theta}{NM} \ln \left(\frac{\frac{\beta_i}{A^2} - \frac{\gamma_i^2}{A^4}}{Vf^i} \right) \frac{\left(\hat{f}_i - \frac{\gamma_i}{A^2} \right)}{\left(\beta_i - \frac{\gamma_i^2}{A^2} \right)}$$

$$\Delta \hat{f}_i = G \left(-\frac{\delta E}{\delta \hat{f}_i} \right) \text{ where } G(u) = \begin{cases} 1, & u > 0 \\ 0, & u = 0 \\ -1, & u < 0 \end{cases}$$

$$V_{new}^i = V_{old}^i + \frac{\left(\Delta \hat{f}_i\right)^2}{A^2} + \frac{2\hat{f}_i \Delta \hat{f}_i}{A^2} - \frac{2\gamma_i^{old} \Delta \hat{f}_i}{A^4} - \frac{\left(\Delta \hat{f}_i\right)^2}{A^4}$$

$$\Delta E = -\frac{1}{2} w_{ii} \left(\Delta \hat{f}_i \right)^2 - u_i \Delta \hat{f}_i + \frac{\theta}{NM} \left(\ln \left[\frac{V_{new}^i}{Vf^i} \right]^2 - \ln \left[\frac{V_{old}^i}{Vf^i} \right]^2 \right)$$

  repeat
  {

$$\hat{f}_i(t+1) = K(\hat{f}_i(t) + \Delta \hat{f}_i)$$

$$\text{where } K(u) = \begin{cases} 0, & u < 0 \\ u, & 0 \le u \le S \\ S, & u > S \end{cases}$$

$$u_i = u_i + w_{ii} \Delta \hat{f}_i$$

$$V_{old}^i = V_{new}^i$$
$$\gamma_i^{old} = \gamma_i^{old} + \Delta \hat{f}_i$$
$$V_{new}^i = V_{old}^i + \frac{\left(\Delta \hat{f}_i\right)^2}{A^2} + \frac{2\hat{f}_i \Delta \hat{f}_i}{A^2} - \frac{2\gamma_i^{old}\Delta \hat{f}_i}{A^4} - \frac{\left(\Delta \hat{f}_i\right)^2}{A^4}$$
$$\Delta E = -\frac{1}{2}w_{ii}\left(\Delta \hat{f}_i\right)^2 - u_i \Delta \hat{f}_i + \frac{\theta}{NM}\left(\ln\left[\frac{V_{new}^i}{Vf^i}\right]^2 - \ln\left[\frac{V_{old}^i}{Vf^i}\right]^2\right)$$

until $\Delta E \geq 0$

    }
    $t = t + 1$
}
until $\left(\hat{f}_i(t) = \hat{f}_i(t-1) \forall i = 1, \ldots, L\right)$

Note that Algorithm 4.2 is almost identical to Algorithm 4.1 and still utilizes some features of Algorithm 2.3, specifically the use of bias inputs and interconnection strength matrices.

4.4.2   Analysis

As in the previous section, we will verify the correct operation of Algorithm 4.2. To verify this, we must examine the Log LVR term in (4.19) more closely.

The gradient of the Log LVR term of the cost function when all pixels except $\hat{f}_i$ are held constant is given by:

$$\frac{\delta}{\delta \hat{f}_i} \text{Log LVR} = \frac{4}{NM} \ln\left(\frac{\frac{\beta_i}{A^2} - \frac{\gamma_i^2}{A^4}}{Vf^i}\right)\left(\frac{\frac{\hat{f}_i}{A^2} - \frac{\gamma_i}{A^4}}{\frac{\beta_i}{A^2} - \frac{\gamma_i^2}{A^4}}\right) \tag{4.22}$$

Note that $\beta_i = \sum_{j=1}^{L}(f_j^i)^2$ and $\gamma_i = \sum_{j=1}^{L} f_j^i$ can be rewritten as $\beta_i = \acute{\beta}_i + \hat{f}_i^2$ and $\gamma_i = \acute{\gamma}_i + \hat{f}_i$ where $\acute{\beta}_i = \sum_{j=1,j\neq i}^{L}(f_j^i)^2$ and $\acute{\gamma}_i = \sum_{j=1,j\neq i}^{L} f_j^i$. In this way, we can extract the elements of (4.22) that depend on $\hat{f}_i$. Hence we obtain:

$$\frac{\delta}{\delta \hat{f}_i} \text{Log LVR} =$$

$$\frac{4}{NM} \frac{\left(\ln\left[\hat{f}_i^2(\frac{1}{A^2} - \frac{1}{A^4}) - \frac{2\hat{f}_i\acute{\gamma}_i}{A^4} + \frac{\acute{\beta}_i}{A^2} - \frac{\acute{\gamma}_i^2}{A^4}\right] - \ln(Vf^i)\right)\left(\frac{\hat{f}_i}{A^2} - \frac{\acute{\gamma}_i}{A^4} - \frac{\hat{f}_i}{A^4}\right)}{\hat{f}_i^2(\frac{1}{A^2} - \frac{1}{A^4}) - \frac{2\hat{f}_i\acute{\gamma}_i}{A^4} + \frac{\acute{\beta}_i}{A^2} - \frac{\acute{\gamma}_i^2}{A^4}}$$

If we assume that A is large enough so that $\frac{1}{A^2} \gg \frac{1}{A^4}$, then:

$$\frac{\delta}{\delta \hat{f}_i} \text{Log LVR} =$$

$$\frac{4}{NMA^2} \frac{\left(\ln\left[\dfrac{\hat{f}_i^2}{A^2} - \dfrac{2\hat{f}_i\acute{\gamma}_i}{A^4} + \dfrac{\acute{\beta}_i}{A^2} - \dfrac{\acute{\gamma}_i^2}{A^4}\right] - \ln(Vf^i)\right)\left(\hat{f}_i\left[1 - \dfrac{1}{A^2}\right] - \dfrac{\acute{\gamma}_i}{A^2}\right)}{\dfrac{\hat{f}_i^2}{A^2} - \dfrac{2\hat{f}_i\acute{\gamma}_i}{A^4} + \dfrac{\acute{\beta}_i}{A^2} - \dfrac{\acute{\gamma}_i^2}{A^4}}$$

Observe that $\frac{\acute{\beta}_i}{A^2} - \frac{\acute{\gamma}_i^2}{A^4}$ is an approximation to the local variance at pixel i neglecting the contribution of the value of pixel i itself. Let $\acute{V} = \frac{\acute{\beta}_i}{A^2} - \frac{\acute{\gamma}_i^2}{A^4}$. As A increases in value, $\acute{V}$ approaches the value of the local variance at pixel i. Similarly, we can define an approximation to the local mean of pixel i as:

$$\acute{M} = \frac{1}{A^2}\sum_{j=1,j\neq i}^{NM} \hat{f}^i_j = \frac{\acute{\gamma}_i}{A^2}$$

This leaves us with:

$$\frac{\delta}{\delta\hat{f}_i} \text{ Log LVR} =$$

$$\frac{4}{NMA^2} \frac{\left(\ln\left[\dfrac{\hat{f}_i^2}{A^2} - \dfrac{2\hat{f}_i\acute{M}}{A^2} + \acute{V}\right] - \ln(Vf^i)\right)\left(\hat{f}_i\left[1 - \dfrac{1}{A^2}\right] - \acute{M}\right)}{\dfrac{\hat{f}_i^2}{A^2} - \dfrac{2\hat{f}_i\acute{M}}{A^2} + \acute{V}} \qquad (4.23)$$

There are some points for which (4.23) is undefined. These points are given by:

$$\hat{f}_i \approx \acute{M} \pm A\sqrt{\frac{\acute{M}^2}{A^2} - \acute{V}}$$

At these points the function will have an infinitely negative gradient. In the event that $\acute{V} > \frac{\acute{M}^2}{A^2}$ then the function will be defined for all values of $\hat{f}_i$. Fortunately this will almost always happen. This can be seen by examining the condition for undefined points to exist:

$$\acute{V} \leq \frac{\acute{M}^2}{A^2}$$

Using the formulas for $\acute{V}$ and $\acute{M}$ we get:

$$\frac{\acute{\beta}_i}{A^2} - \frac{\acute{\gamma}_i^2}{A^4} \leq \frac{1}{A^2}\left(\frac{\acute{\gamma}_i}{A^2}\right)^2$$

$$\frac{\acute{\beta}_i}{A^2} - \frac{\acute{\gamma}_i^2}{A^4} \leq \frac{1}{A^6}\acute{\gamma}_i^2$$

Hence we require that:

$$\frac{\acute{\beta}_i}{A^2} \le \acute{\gamma}_i^2 \left(\frac{1}{A^4} + \frac{1}{A^6} \right) \tag{4.24}$$

However, the variance of a set of numbers is always greater than or equal to zero. $\acute{V}$ is the variance of the local region of pixel i obtained when the value of pixel i is set to zero. This gives us:

$$\acute{V} = \frac{\acute{\beta}_i}{A^2} - \frac{\acute{\gamma}_i^2}{A^4} > 0$$

Which means that:

$$\frac{\acute{\beta}_i}{A^2} \ge \frac{\acute{\gamma}_i^2}{A^4} \tag{4.25}$$

Equation (4.25) of course means that condition (4.24) will only be satisfied when the local variance is very close to zero. As long as steps are taken to ensure that this does not occur, the function will be well defined for all values of $\hat{f}_i$.

The points for which equation (4.23) is equal to zero are the stationary points of the Log LVR term in (4.19). Equation (4.23) is zero when $\hat{f}_i = \frac{\acute{M}}{1 - \frac{1}{A^2}}$. This corresponds to the case where $\hat{f}_i$ is approximately equal to its local mean. Equation (4.23) also has zeroes when $\hat{f}_i$ satisfies:

$$\ln \left(\frac{\hat{f}_i^2}{A^2} - \frac{2\hat{f}_i \acute{M}}{A^2} + \acute{V} \right) - \ln(Vf^i) = 0$$

This is equivalent to:

$$\frac{\hat{f}_i^2}{A^2} - \frac{2\hat{f}_i \acute{M}}{A^2} + J = 0$$

where $J = \acute{V} - Vf^i$.

The stationary points are thus given by:

$$\hat{f}_i \approx \acute{M} \pm A \sqrt{\frac{\acute{M}^2}{A^2} - J} \tag{4.26}$$

Note that (4.26) indicates that if $J > 0$, (4.26) may not have a real-valued solution. Equation (4.26) is identical to (4.18) in the previous section and so the case by case analysis in Section 4.3 is identical for a Log LVR-modified cost function as it was for the LVMSE-modified cost function.

As with the LVMSE term, the Log LVR term in the cost function is well behaved despite its non-linear nature. In all cases minimum exist and by minimizing the Log LVR term in the cost function, the local variance of the current pixel's region will always move closer to the target variance estimate.

4.5   Implementation Considerations

A problem to be overcome is that the third terms in the LVMSE-based cost functions are not quadratic in nature. When the local variance in the image estimate is much lower than the projected local variances of the original image, the LVMSE term in Algorithm 4.1 becomes large and may force the pixel values to an extreme of the range of acceptable values in order to create a high variance region. The LVMSE term should never completely dominate over the first term in (4.2) since the LVMSE term only attempts to match regions, not pixels, and fine structure within the region will be lost. To remedy this situation, the pixel values are not allowed to change by more than a set amount per iteration. This method appears to work well in practice and the pixel values converge to a solution after a finite number of iterations. This method, however, is not required to the same degree in Algorithm 4.2. Algorithm 4.2 was designed to avoid this effect; however, this method may still be employed to improve results.

The addition of the LVMSE term into the cost function allows a powerful optimization to be made to Algorithm 4.1. In regions where the degraded image is very smooth and the variance estimate of the original image is very small, improvement in image processing speed can be achieved by not restoring these pixels. This will not affect the quality of processing since attempting to deconvolve regions where the blurring effect is not noticeable by humans can only serve to amplify noise. It is logical not to attempt to restore such regions when using Algorithm 4.1 since the LVMSE-based term in the cost function for this algorithm has little effect at low variance regions. Algorithm 4.2 on the other hand was designed to smooth these regions and so it is not necessary to avoid attempting to restore these regions.

4.6   Numerical Examples

In this section, a number of examples are given to show the performance of the methods we examined in this chapter. Comparisons will be made with some well-known methods in the literature.

4.6.1   Color Image Restoration

For the first experiment, color images were used consisting of three color planes, red, green and blue. The image was degraded by a 5 by 5 Gaussian PSF of standard deviation 2.0 applied to each of the color planes. In addition, additive noise of variance 369.31 was also added to each color plane. Figure 4.4a shows the original image and Figure 4.4b shows the degraded image. The degraded image has an SNR of 19.81 dB and an LSMSE of 313.05. The SNR was calculated by adding together the signal to noise ratio of each color plane:

$$ \text{SNR} = 20 \log \left(\frac{\sigma_o^r}{\sigma_n^r} + \frac{\sigma_o^g}{\sigma_n^g} + \frac{\sigma_o^b}{\sigma_n^b} \right) \tag{4.27} $$

Similarly the LSMSE for the entire image was calculated by summing the LSMSEs of each color plane. A 9 by 9 neighborhood was used for calculating the local variance. In this example, we assumed that each color plane in our test image does not have a high level of correlation and so the filters are applied to each color plane separately. A Wiener filter restored image is shown in Figure 4.4c and has an SNR of 16.65 dB and an LSMSE of 859.80 [4, 21]. The image was also restored using Algorithm 2.3, without the LSMSE term. A constraint factor of $\lambda = 0.001$ was chosen. The CLS restored image is shown in Figure 4.4d and has an SNR of 17.26 dB and an LSMSE of 634.04. The image was also restored using the variance selection adaptive constraint algorithm from Chapter 3. This image is shown in Figure 4.4e and has a SNR of 19.19 dB and a LSMSE of 195.68. The same degraded image was also restored using the LSMSE-modified cost function, Algorithm 4.1. In the LSMSE-modified cost function, the value of λ was set to 0.0005. The factor θ was set to be 0.00001 and the image local variance estimate was computed as:

$$\sigma_A^2(\mathbf{g}(x,y))^\star = 2\left(\sigma_A^2(\mathbf{g}(x,y)) - 200\right)$$

This image is shown in Figure 4.4f and has an SNR of 19.89 dB and an LSMSE of 180.81. Finally the degraded image was restored using the Log LVR-modified cost function, Algorithm 4.2. In the Log LVR-modified cost function, the value of λ was set to 0.0005. The factor θ was set to be 50 and the image local variance estimate was computed as for Algorithm 4.1. This image is shown in Figure 4.4g and has an SNR of 21.65 dB and an LSMSE of 88.43.

By visual observation it can be seen that Figures 4.4f and 4.4g, produced by the LSMSE and Log LVR-based cost functions, display better noise suppression in background regions and are at the same time sharper than Figures 4.4c and 4.4d, produced by the Wiener and the CLS approaches. Figures 4.4f and 4.4g also display a better SNR and LSMSE than Figures 4.4c, 4.4d and 4.4e. Although the LSMSE restored images are visually closer to the original image than the degraded image, their SNRs are only slightly higher than the degraded image. This is not surprising in view of the arguments above that SNR does not correspond well with human visual perception. However, LSMSE does match with human observation and assigns a much lower value to Figures 4.4f and 4.4g. Comparing the two different forms of the LSMSE-based cost functions, we find that Algorithm 4.2 (Figure 4.4g) is superior, with a similar level of sharpness when compared to Figure 4.4f, yet better noise suppression in background regions.

We see that the variance selection adaptive constraint method produces a similar result to Algorithm 4.1. This is primarily because both algorithms use the concept of a variance threshold. As mentioned in Section 4.5, if the local variance is below the threshold, the pixel is not adjusted. Both Algorithms 2.3 and 4.1 use identical thresholds and so have similar LSMSEs. Algorithm 4.2, however, was designed not to require the variance threshold and instead provides additional smoothing to background regions and hence a much lower LSMSE.

Figure 4.4: Color images restored using various algorithms
(grayscale versions shown in this book)

4.6.2 Grayscale Image Restoration

For the second example, a grayscale image was degraded by a 5 by 5 Gaussian PSF of standard deviation 2.0. Additive noise of variance 87.62 was also added. Figure 4.5a shows the original image and Figure 4.5b shows the degraded image. The degraded image has an SNR of 12.58 dB and an LSMSE of 28.13. The degraded image was first restored using a Wiener filter approach. The Wiener restored image is shown in Figure 4.5c and has an SNR of 11.66 dB and an LSMSE of 38.69. The image was also restored using the CLS algorithm (Algorithm 2.3). Figure 4.5d shows the image restored using the CLS algorithm with a constant factor of $\lambda = 0.001$. Figure 4.5d has an SNR of 8.76 dB and an LSMSE of 128.09. Figure 4.5e shows the image restored using the CLS algorithm with a constant factor of $\lambda = 0.002$. Figure 4.5e has an SNR of 11.93 dB and an LSMSE of 36.91. Figure 4.5f shows the image restored using the adaptive constraint algorithm presented in Chapter 3 using a range of constraint values from 0.02 to 0.0015 associated with levels of local variance. Figure 4.5f has an SNR of 11.97 dB and an LSMSE of 22.28. The degraded image was also restored using the LVMSE-modified cost function implemented using Algorithm 4.1. Figure 4.5g shows this image which has an SNR of 12.15dB and an LSMSE of 22.71. Finally the degraded image was restored using the Log LVR-modified cost function implemented using Algorithm 4.2. Figure 4.5h shows this image which has an SNR of 12.07 dB and an LSMSE of 20.59. By observation, it can be seen that Figure 4.5h is visually closest to the original image. LSMSE confirms visual inspection and indicates that Figure 4.5h is the most well restored. Note that once again the adaptive algorithm from Chapter 3 performs similarly to the LSMSE-based algorithms. The advantage of the LSMSE algorithms is that they have fewer free variables to set up.

4.6.3 LSMSE of Different Algorithms

For the third example, the original flower image was blurred using a 5 by 5 Gaussian blur of standard deviation 2.0. A number of images were created, each suffering a different value of noise. The images were restored using Algorithm 2.3, Algorithm 4.1, Algorithm 4.2, a Wiener filter and the adaptive constraint algorithm from Chapter 3. For each image, the same value of λ was used in Algorithm 2.3, Algorithm 4.1 and Algorithm 4.2. This meant that the restored images from Algorithm 2.3, Algorithm 4.1 and Algorithm 4.2 had the same degree of sharpness, but differed in the level of noise suppression. In this way the effects of the LSMSE-based terms in (4.2) and (4.19) could be examined in isolation. Figure 4.6 shows the results of this experiment. It can be clearly seen that in terms of the LSMSE, Algorithms 4.1 and 4.2 outperform the other algorithms, especially the standard CLS approach for the same level of sharpness.

4.6.4 Robustness Evaluation

For the fourth example, the original flower image was blurred using a 5 by 5 Gaussian blur of standard deviation 2.0. Additive noise of variance 87.62 was

Figure 4.5: Grayscale images restored using various algorithms

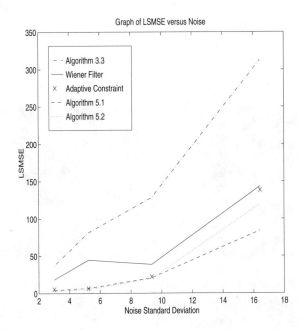

Figure 4.6: Graph of LSMSE for various algorithms and levels of noise

also added. The degraded image was restored using Algorithm 2.3, Algorithm 4.1 and Algorithm 4.2. In each algorithm, λ was set to 0.001 to maintain the same level of sharpness. Figure 4.7a shows the results of Algorithm 2.3. This image is identical to Figure 4.5d and has an SNR of 8.76 dB and an LSMSE of 128.09. Figure 4.7b shows the results of Algorithm 4.1. Figure 4.7b has an SNR of 12.45 dB and an LSMSE of 20.76. Figure 4.7c shows the results of Algorithm 4.2. Figure 4.7c has an SNR of 12.25 dB and an LSMSE of 19.76. Next we severed one of the neural interconnections to a neighboring neuron for every neuron in the network. The same connection was severed for each neuron in the network. This would be expected to degrade the performance of the network. Using the same parameters, the restorations were performed again. Figure 4.7d shows the results of restoring the image using Algorithm 2.3 with a faulty network. The SNR is minus 4.21 dB and the LSMSE is 5889.42. Figure 4.7e shows the results of restoring the image using Algorithm 4.1 with a faulty network. The SNR is 12.15 dB and the LSMSE is 23.23. Figure 4.7f shows the results of restoring the image using Algorithm 4.2 with a faulty network. The SNR is 10.06 dB and the LSMSE is 40.13. From these results we can see that Algorithm 2.3 is not very tolerate of errors in weights. The image produced by the faulty network is very degraded and has poor values of SNR and LSMSE. On the other hand, Algorithm 4.1 and Algorithm 4.2 have almost no visual differences between images restored using the correct network and images restored using the faulty network. The images restored using the faulty network have only slightly worse values of SNR and LSMSE compared to the image restored using the correct network. The

Figure 4.7: Images restored using correct and faulty networks

reason that Algorithm 4.1 and 4.2 are more fault-tolerant than Algorithm 2.3 is due to the LSMSE-related terms in these algorithms. The damaged weights in Algorithm 2.3 produced streaks in the image. These streaks would cause the pixels in their vicinity to have very high local variances. Since Algorithm 2.3 does not consider the local regional statistics of the image, the streaks are not suppressed. However, Algorithm 4.1 and Algorithm 4.2 attempt to match local variances in the restored image with an estimate of the original image. The streaks are therefore suppressed by Algorithms 4.1 and 4.2. It is clear that Algorithms 4.1 and 4.2 are very robust and are not greatly affected by errors in the network. Algorithm 4.1 is more robust than Algorithm 4.2 on the edges because of the fact that the log-ratio relationship between the local variance and the target variance used by Algorithm 4.2 was developed to de-emphasize the LSMSE effect on edges. Algorithm 4.1 has its greatest effect on edges, whereas Algorithm 4.2 was specifically designed to have the least effect on edges and the greatest effect on smooth regions. However, both algorithms are still quite tolerate of network errors. This is because the LVMSE-based terms in these algorithms can never be affected by severed neural connections.

4.7 Summary

In Chapter 1, a novel error measure was introduced which compares two images by consideration of their regional statistical differences rather than their pixel-level differences. It was found that this error measure more closely corresponds to human visual perception of image quality. Based on the new error measure, two cost functions were presented in this chapter. The first cost function was based closely on the LVMSE error measure introduced in Chapter 1. This cost function was analyzed and shown to be well behaved. The analysis of the first modified cost function suggested that improvements could be made by incorporating a logarithmic version of the LVMSE into the standard cost function. The second cost function was hence introduced and shown to be well behaved. Algorithms to optimize these cost functions were designed based on adaptation of the neural network approach to optimizing constrained least square error.

The introduced algorithms were shown to suppress noise strongly in low variance regions while still preserving edges and highly textured regions of an image. The algorithms were shown to perform well when applied to both grayscale and color images. It was also shown that the proposed iterative algorithms are very robust.

Chapter 5

Model-Based Adaptive Image Restoration

5.1 Model-Based Neural Network

In this chapter, we introduce the technique of model-based neural networks, which is to be applied to our problem of adaptive regularization in image restoration. Instead of adopting the general neural network architecture introduced in Chapter 1 for our adaptive image processing applications, we propose the use of *modular model-based* neural network for our purpose. We used the term "model-based" in the sense of Caelli *et al.* [62], where expert knowledge in the problem domain is explicitly incorporated into a neural network by restricting the domain of the network weights to a suitable subspace in which the solution of the problem resides. In this way, a single weight vector can be uniquely specified by a small number of parameters and the "curse of dimensionality" problem is partially alleviated.

To appreciate this formulation more readily, we review some fundamental concepts of artificial neuron computation, where each such neuron is the elementary unit of computation in a neural network [34, 35]. In general, the sth neuron in the network implements a mapping $f_s : \mathbf{R}^N \longrightarrow \mathbf{R}$ which is given by

$$
\begin{aligned}
y_s &= f_s(\mathbf{x}) \\
&= g(\mathbf{p}_s^T \mathbf{x}) \\
&= g\left(\sum_{n=1}^{N} p_{qs} x_n \right)
\end{aligned}
\tag{5.1}
$$

where

$$
\begin{aligned}
\mathbf{x} &= [x_1, \ldots, x_N]^T \in \mathbf{R}^N \quad \text{and} \\
\mathbf{p}_s &= [p_{s1}, \ldots, p_{sN}]^T \in \mathbf{R}^N
\end{aligned}
$$

111

are the input vector and the weight vector for the neuron, respectively. g is usually a non-linear sigmoid function which limits the output dynamic range of the neuron. We will extend this concept and define model-based neuron in the next section.

5.1.1   Weight-Parameterized Model-Based Neuron

The main assumption in this weight-parameterized model-based formulation [62] is that for a specific domain of knowledge, the corresponding weight domain is restricted to a low-dimensional submanifold of $\mathbf{R}^N$.

Denoting this weight domain by $\mathcal{W}_p$, the formulation thus assumes the existence of a mapping $\mathcal{M} : \mathbf{R}^M \longrightarrow \mathcal{W}_p \subset \mathbf{R}^N$ such that

$$\mathbf{p} = \mathcal{M}(\mathbf{z}) \tag{5.2}$$

where

$$\mathbf{z} = [z_1, \ldots, z_M]^T \in \mathbf{R}^M$$
$$\mathbf{p} = [p_1, \ldots, p_N]^T \in \mathbf{R}^N$$
$$= [\mathcal{M}_1(\mathbf{z}), \ldots, \mathcal{M}_N(\mathbf{z})]^T$$

with $M < N$. The mappings $\mathcal{M}_n : \mathbf{R}^M \longrightarrow \mathbf{R}, n = 1, \ldots, N$ are the component functions of $\mathcal{M}$. The structure of a typical weight-parameterized model-based neuron is shown in Figure 5.1.

Assuming that each component function is differentiable with respect to $\mathbf{z}$,

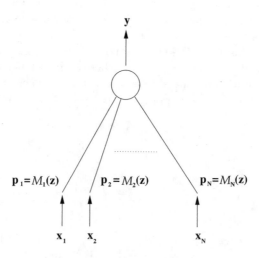

Figure 5.1: The weight-parameterized model-based neuron

the steepest descent update rule for the components of $\mathbf{z}$ is as follows

$$
\begin{aligned}
z_m(t+1) &= z_m(t) - \eta\, \frac{\partial \mathcal{E}_t}{\partial z_m} \\
&= z_m(t) - \eta \sum_{n=1}^{N} \frac{\partial \mathcal{E}_t}{\partial p_n} \frac{\partial p_n}{\partial z_m}
\end{aligned}
\tag{5.3}
$$

where $\mathcal{E}_t$ is an instantaneous error measure between the network output and the desired output.

As a result, if we possess prior knowledge of our problem domain in the form of the mapping $\mathcal{M}$ and if $M \ll N$, the optimization can proceed within a subspace of greatly reduced dimensionality and the problem of "curse of dimensionality" is partially alleviated. In the next section we will present an alternative formulation of the adaptive regularization problem in terms of this model-based neuron.

5.2 Hierarchical Neural Network Architecture

While the previous model-based formulation describes the operations of individual neurons, the term "modular neural network" [47, 119, 120, 121, 122, 123, 124] refers to the overall architecture where specific computations are localized within certain structures known as *sub-networks*. This class of networks serves as natural representations of training data arising from several distinct classes, where each such class is represented by a single sub-network. Specific implementations of modular neural networks include the various hierarchical network architecture described in [48, 125, 126, 127, 128].

We are especially interested in the particular hierarchical implementation of the modular network structure by Kung and Taur [48]. Their proposed architecture associates a *sub-network output* $\phi(\mathbf{x}, \mathbf{w}_r)$ with the rth sub-network. In addition, we define lower-level *neurons* within each sub-network with their corresponding *local neuron output* $\psi_r(\mathbf{x}, \mathbf{w}_{s_r})$, $s_r = 1, \ldots, S_r$. The sub-network output is defined as the linear combination of the local neuron outputs as follows:

$$
\phi(\mathbf{x}, \mathbf{w}_r) = \sum_{s_r=1}^{S_r} c_{s_r} \psi_r(\mathbf{x}, \mathbf{w}_{s_r})
\tag{5.4}
$$

where c_{s_r} are the combination coefficients. The hierarchical architecture and the structure of its sub-network is shown in Figure 5.2.

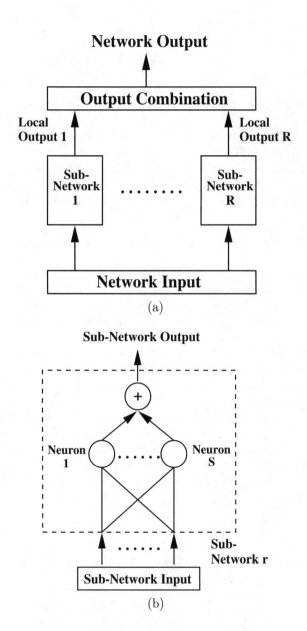

Figure 5.2: Hidden-node hierarchical network architecture (a) global network architecture (b) sub-network architecture

5.3 Model-Based Neural Network with Hierarchical Architecture (HMBNN)

Given the previous description of the model-based neuron, which specifies the computational operations at the neuronal level, and the overall hierarchical structure, which specifies the macroscopic network architecture, it is natural to combine the two in a single framework. More specifically, we can incorporate model-based neuron computation at the lowest hierarchy, i.e., at the level of single neuron within a sub-network, of the hierarchical neural network.

Formally, if we assume $\mathbf{w}_{s_r} \in \mathbf{R}^N$, the local neuron output $\psi_r(\mathbf{x}, \mathbf{w}_{s_r})$ can be specified as follows:

$$\psi_r(\mathbf{x}, \mathbf{w}_{s_r}) \equiv \psi_r(\mathbf{x}, \mathcal{M}(\mathbf{z}_{s_r})) \qquad (5.5)$$

where $\mathbf{z}_{s_r} \in \mathbf{R}^M$, with $M < N$, is the model-based weight vector, and $\mathcal{M} : \mathbf{R}^M \longrightarrow \mathbf{R}^N$ is the mapping relating the model-based weight vector to the neuronal weight vector. In this case, the lower-dimensional model-based vector $\mathbf{z}_{s_r}$ is embedded in the higher-dimensional weight vector $\mathbf{w}_{s_r}$, and the weight optimization is carried out in the lower-dimensional weight space $\mathbf{R}^M$.

5.4 HMBNN for Adaptive Image Processing

Our current formulation of the adaptive regularization problem requires the partition of an image into disjoint regions, and the assignment of the optimal regularization parameter value λ to the respective regions. It is natural, therefore, to assign a single sub-network to each such region, and regard the regularization parameter λ as a model-based weight to be optimized using a special set of training examples. More specifically, for the rth sub-network, it will be shown in the next section that the image restoration process is characterized by the evaluation of the local neuron output as a linear operation as follows.

$$\psi_r(\mathbf{x}, \mathbf{p}_{s_r}) = \mathbf{p}_{s_r}^T \mathbf{x} \qquad (5.6)$$

where $\mathbf{x} \in \mathbf{R}^N$ denotes a vector of image gray level values in a local neighborhood and $\mathbf{p}_{s_r}$ is the *image restoration convolution mask* derived from the point spread function (PSF) of the degradation mechanism. The dimension N of the vectors $\mathbf{x}$ and $\mathbf{p}_{s_r}$ depends on the size of the PSF. It will be shown that the convolution mask coefficient vector $\mathbf{p}_{s_r}$ can be expressed as a function of the regularization parameter λ. In other words, there exists a mapping $\mathcal{M} : \mathbf{R} \longrightarrow \mathbf{R}^N$ such that

$$\mathbf{p}_{s_r} = \mathcal{M}(\lambda) \qquad (5.7)$$

which is equivalent to the embedding of the scalar parameter λ in the high-dimensional space $\mathbf{R}^N$, and thus corresponds to a weight-parameterized model-based neuron.

5.5   The Hopfield Neural Network Model for Image Restoration

We recall that, in regularized image restoration, the cost function consists of a data-conformance evaluation term and a model-conformance evaluation term [24, 26]. The model-conformance term is usually specified as a continuity constraint on neighboring gray level values in image processing applications. The contribution of these two terms is adjusted by the so-called regularization parameter or the Lagrange multiplier which allows the two terms to combine additively:

$$E = \frac{1}{2}\|\mathbf{y} - \mathbf{H}\hat{\mathbf{f}}\|^2 + \frac{1}{2}\lambda\|\mathbf{D}\hat{\mathbf{f}}\|^2 \tag{5.8}$$

where the vector $\mathbf{y}$, with $y_i, i = 1, \ldots, N_I$ as components and N_I as the number of image pixels, denotes the blurred image with its pixel values lexicographically ordered. The vectors $\hat{\mathbf{f}}$ with components $\hat{f}_i$ is the corresponding restored image and the matrix $\mathbf{H}$, with components $h_{ij}, i, j = 1, \ldots, N_I$, is the blur function. The matrix $\mathbf{D}$ with components d_{ij} is a linear operator on $\hat{\mathbf{f}}$ and λ is the regularization parameter. For image restoration purpose, $\mathbf{D}$ is usually a differential operator, and the minimization of the above cost function effectively limits the local variations of the restored image.

In the current work, the optimization of the primary image restoration cost function E was performed within the framework of neural network optimization where a modified neural network architecture based on the Hopfield neural network model in Chapter 2 was employed. This *primary* Hopfield image restoration neural network, which is mainly responsible for optimizing the value of E, is to be carefully distinguished from the HMBNN described previously, which serves as a *secondary* neural network to optimize the regularization parameter λ in E.

It was shown in Chapter 2 that the neuron output can be updated using the following equation:

$$\triangle \hat{f}_i = -\frac{u_i(t)}{w_{ii}} \tag{5.9}$$

and the final output of the neuron is given by

$$\hat{f}_i(t + 1) = \hat{f}_i(t) + \triangle \hat{f}_i(t) \tag{5.10}$$

We can see that equation (5.9) expresses the change in neuron output in terms of w_{ii}, which is in turn a function of the regularization parameter λ. This dependency is essential for the re-formulation of the current adaptive regularization problem into a learning problem for a set of model-based neurons.

5.6   Adaptive Regularization: An Alternative Formulation

In this section we propose an alternative formulation of adaptive regularization which centers on the concept of *regularization parameters as model-based neuronal weights*. By adopting this alternative viewpoint, a neat correspondence is found to exist between the mechanism of adaptive regularization and the computation of an artificial model-based neuron.

Referring to equation (5.9), we can express the instantaneous restoration stepsize $\triangle \hat{f}_i(t)$ in (5.9) in the form of a neural computational process as follows

$$
\begin{aligned}
\triangle \hat{f}_i &= -\frac{u_i(t)}{w_{ii}} \\
&= -\frac{\sum_{j=1}^{N_I} w_{ij}\hat{f}_j + b_i}{w_{ii}} \\
&= \sum_{j=1}^{N_I} p_{ij}\hat{f}_j + q_i b_i \\
&= \mathbf{p}_i^T \hat{\mathbf{f}}_i
\end{aligned}
\tag{5.11}
$$

where

$$
\mathbf{p}_i = [p_{i1}, \ldots, p_{iN_I}, q_i]^T \in \mathbf{R}^{N_I+1}
$$
$$
\hat{\mathbf{f}}_i = [\hat{f}_1, \ldots, \hat{f}_{N_I}, b_i]^T \in \mathbf{R}^{N_I+1}
$$

and the weights of this hypothetical neuron are defined as

$$
p_{ij} = -\frac{w_{ij}}{w_{ii}}
\tag{5.12}
$$

$$
q_i = -\frac{1}{w_{ii}}
\tag{5.13}
$$

The weights w_{ij} of the primary Hopfield neuron (as opposed to those of this hypothetical neuron) are in turn given by the following equation (see Chapter 2):

$$
\begin{aligned}
w_{ij} &= -\sum_{p=1}^{N_I} h_{pi}h_{pj} - \lambda \sum_{p=1}^{N_I} d_{pi}d_{pj} \\
&= g_{ij} + \lambda l_{ij}
\end{aligned}
\tag{5.14}
$$

where we define

$$
g_{ij} = -\sum_{p=1}^{N_I} h_{pi}h_{pj}
\tag{5.15}
$$

$$
l_{ij} = -\sum_{p=1}^{N_I} d_{pi}d_{pj}
\tag{5.16}
$$

From these equations, the weights p_{ij} of the hypothetical neuron can be expressed as a function of the regularization parameter λ as follows

$$
p_{ij} = -\frac{g_{ij} + \lambda l_{ij}}{g_{ii} + \lambda l_{ii}}
\tag{5.17}
$$

$$
q_i = -\frac{1}{g_{ii} + \lambda l_{ii}}
\tag{5.18}
$$

To re-interpret these equations as the computation of a model-based neuron, we re-cast them into its two-dimensional form from the previous lexicographical ordered form. Defining the *lexicographical mapping* $\mathcal{L}$ as follows

$$i = \mathcal{L}(i_1, i_2) = i_1 N_x + i_2 \tag{5.19}$$

where (i_1, i_2) is the corresponding 2D position of the ith lexicographical vector entry in the image lattice, and N_x is the width of the image lattice. With this mapping, we can re-interpret equation (5.11) in a 2D setting

$$\triangle \widetilde{f}_{i_1,i_2} = \sum_{k=-L_c}^{L_c} \sum_{l=-L_c}^{L_c} \widetilde{p}_{i_1,i_2,k,l} \widetilde{f}_{i_1+k,i_2+l} + \widetilde{q}_{i_1,i_2} \widetilde{b}_{i_1,i_2} \tag{5.20}$$

where the tilded form of the variables indicate that the current quantity is indexed by its 2D position in the lattice, rather than its 1D position in the lexicographically ordered vector. The summation in the equation is taken over the support of a *2D neuronal weight mask*, the size of which depends on the extent of the original point spread function [46] and which is $(2L_c + 1)^2$ in this case. For $i = \mathcal{L}(i_1, i_2)$, we define

$$\widetilde{f}_{i_1,i_2} = \hat{f}_i \tag{5.21}$$

$$\widetilde{p}_{i_1,i_2,k,l} = p_{i,\mathcal{L}(i_1+k,i_2+l)} \tag{5.22}$$

and the variables $\triangle \widetilde{f}_{i_1,i_2}$, $\widetilde{q}_{i_1,i_2}$ and $\widetilde{b}_{i_1,i_2}$ are similarly defined as $\widetilde{f}_{i_1,i_2}$. For spatially invariant degradation, the variables $\widetilde{p}_{i_1,i_2,k,l}$ and $\widetilde{q}_{i_1,i_2}$ are independent of the position (i_1, i_2) in the image lattice, and the above equation can be re-written as

$$\triangle \widetilde{f}_{i_1,i_2} = \sum_{k=-L_c}^{L_c} \sum_{l=-L_c}^{L_c} \widetilde{p}_{k,l} \widetilde{f}_{i_1+k,i_2+l} + \widetilde{q} \widetilde{b}_{i_1,i_2}$$

$$= \widetilde{\mathbf{p}}^T \widetilde{\mathbf{f}}_{i_1,i_2} \tag{5.23}$$

where

$$\widetilde{\mathbf{p}} = [\widetilde{p}_{-L_c,-L_c}, \ldots, \widetilde{p}_{0,0}, \ldots, \widetilde{p}_{L_c,L_c}, \widetilde{q}]^T \in \mathbf{R}^{N_c+1}$$

$$\widetilde{\mathbf{f}}_{i_1,i_2} = [\widetilde{f}_{i_1-L_c,i_2-L_c}, \ldots, \widetilde{f}_{i_1,i_2}, \ldots, \widetilde{f}_{i_1+L_c,i_2+L_c}, \widetilde{b}_{i_1,i_2}]^T \in \mathbf{R}^{N_c+1}$$

and $N_c = (2L_c + 1)^2$.

5.6.1 Correspondence with the General HMBNN Architecture

For spatially invariant degradation with small support, we have the condition $N_c \ll N_I$, where N_I is the number of pixels in the image, and we can view equation (5.23) as a local convolution operation over a selected neighborhood of

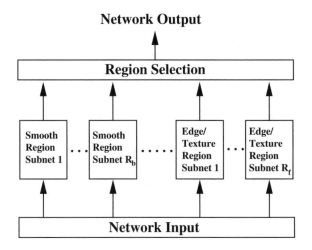

Figure 5.3: The model-based neural network with hierarchical architecture for adaptive regularization

$\widetilde{f}_{i_1,i_2}$. On the other hand, due to the invariant nature of $\widetilde{p}_{k,l}$ and $\widetilde{q}$ for all i_1 and i_2, this operation can alternatively be viewed as the computational process of a model-based neuron with input vector $\widetilde{\mathbf{f}}_{i_1,i_2}$ and weight vector $\widetilde{\mathbf{p}}$. In other words, if we assume that the image is subdivided into regions $\mathcal{R}_r, r = 1, \ldots, R$, and we assign a sub-network with a single model-based neuron to each region, i.e., $S_r = 1$ for each r according to the notation adopted previously, then the local model-based neuron output corresponding to $\mathcal{R}_r$ can be represented as

$$\psi_r(\widetilde{\mathbf{f}}_{i_1,i_2}, \widetilde{\mathbf{p}}_{k,l}(\lambda_r)) = \widetilde{\mathbf{p}}_{k,l}^T(\lambda_r)\widetilde{\mathbf{f}}_{i_1,i_2} \tag{5.24}$$

where λ_r, the regional regularization parameter, can be considered the *scalar* model-based weight associated with each region $\mathcal{R}_r$. As a result it can be classified into the class of hidden-node hierarchical networks. The architecture of this HMBNN for adaptive regularization is shown in Figure 5.3.

Adopting the weight-parameterized model-based neuron structure due to the natural way in which $\lambda_r \in \mathbf{R}$ is embedded in the high-dimensional weight vector $\widetilde{\mathbf{p}}_{k,l}(\lambda_r) \in \mathbf{R}^{N_c+1}$, and referring to equation (5.17), we can obtain the component mappings $\mathcal{M}_n$ of the model-based neuron for the rth sub-network as follows

$$\widetilde{p}_{k,l}(\lambda_r) = \mathcal{M}_n(\lambda_r) = -\frac{\widetilde{g}_{k,l} + \lambda_r \widetilde{l}_{k,l}}{\widetilde{g}_{0,0} + \lambda_r \widetilde{l}_{0,0}} \qquad n = 1, \ldots, N_c \tag{5.25}$$

$$\widetilde{q}(\lambda_r) = \mathcal{M}_{N_c+1}(\lambda_r) = -\frac{1}{\widetilde{g}_{0,0} + \lambda_r \widetilde{l}_{0,0}} \tag{5.26}$$

where the mapping index n corresponds to some specific arrangement of the indices (k, l) in the 2D weight mask $\widetilde{p}_{k,l}$, and the values of $\widetilde{g}_{k,l}$ and $\widetilde{l}_{k,l}$ can be

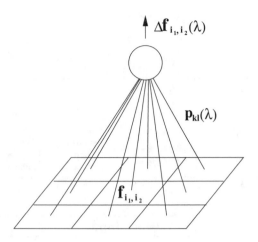

Figure 5.4: The model-based neuron for adaptive regularization

obtained from their lexicographical counterparts g_{ij} and l_{ij} using equation (5.22). The resulting concatenation $\mathcal{M}$ of the components is thus a mapping from $\mathbf{R}$ to $\mathbf{R}^{N_c+1}$. Corresponding to the notation of Section 5.1, we have $M = 1$ and $N = N_c+1$ for the current adaptive regularization problem. In addition, each mapping is in the form of a *first-order rational polynomial* of λ_r, the differentiability of which ensures that the weights $\widetilde{p}_{k,l}$ are trainable. The structure of the model-based neuron adapted for restoration is shown in Figure 5.4.

To complete the formulation, we define a training set for the current model-based neuron. Assuming that for a certain region $\mathcal{R}_r$ in the image lattice, there exists a desired restoration behavior $\triangle \widetilde{f}^d_{i_1,i_2}$ for each pixel $(i_1, i_2) \in \mathcal{R}_r$. We can then define the training set $\mathcal{V}_r$ as follows

$$\mathcal{V}_r = \{(\widetilde{\mathbf{f}}_{i_1,i_2}, \triangle \widetilde{f}^d_{i_1,i_2}) : (i_1, i_2) \in \mathcal{R}_r\} \tag{5.27}$$

where the input vector $\widetilde{\mathbf{f}}_{i_1,i_2}$ contains the gray level values of pixels in the N_c-neighborhood of (i_1, i_2). Defining the instantaneous cost function $\mathcal{E}_t$ of the neuron as follows,

$$\mathcal{E}_t = \frac{1}{2}(\triangle \widetilde{f}^d_{i_1,i_2} - \triangle \widetilde{f}_{i_1,i_2})^2 \tag{5.28}$$

we can then apply steepest descent to modify the value of the regularization parameter λ_r.

$$\lambda_r(t+1) = \lambda_r(t) - \eta \frac{\partial \mathcal{E}_t}{\partial \lambda_r} \tag{5.29}$$

where

$$\frac{\partial \mathcal{E}_t}{\partial \lambda_r} = -(\triangle \widetilde{f}^d_{i_1,i_2} - \triangle \widetilde{f}_{i_1,i_2}) \frac{\partial(\triangle \widetilde{f}_{i_1,i_2})}{\partial \lambda_r} \tag{5.30}$$

To evaluate the partial derivative in equation (5.30), we first define the following two quantities

$$\widetilde{\alpha}_{i_1,i_2} = \sum_{k=-L_c}^{L_c} \sum_{l=-L_c}^{L_c} \widetilde{g}_{k,l} \widetilde{f}_{i_1+k,i_2+l} \tag{5.31}$$

$$\widetilde{\beta}_{i_1,i_2} = \sum_{k=-L_c}^{L_c} \sum_{l=-L_c}^{L_c} \widetilde{l}_{k,l} \widetilde{f}_{i_1+k,i_2+l} \tag{5.32}$$

From equation (5.23), the change in neuron output, $\triangle \widetilde{f}_{i_1,i_2}$, is then expressed in terms of these quantities as follows,

$$
\begin{aligned}
\triangle \widetilde{f}_{i_1,i_2} &= \sum_{k=-L_c}^{L_c} \sum_{l=-L_c}^{L_c} \widetilde{p}_{k,l} \widetilde{f}_{i_1+k,i_2+l} + \widetilde{q} \widetilde{b}_{i_1,i_2} \\
&= - \frac{\sum_{k=-L_c}^{L_c} \sum_{l=-L_c}^{L_c} (\widetilde{g}_{k,l} + \lambda_r \widetilde{l}_{k,l}) \widetilde{f}_{i_1+k,i_2+l} + \widetilde{b}_{i_1,i_2}}{\widetilde{g}_{0,0} + \lambda_r \widetilde{l}_{0,0}} \\
&= - \frac{\widetilde{\alpha}_{i_1,i_2} + \lambda_r \widetilde{\beta}_{i_1,i_2} + \widetilde{b}_{i_1,i_2}}{\widetilde{g}_{0,0} + \lambda_r \widetilde{l}_{0,0}}
\end{aligned}
\tag{5.33}
$$

From this equation, we can evaluate the derivative in equation (5.30)

$$
\begin{aligned}
\frac{\partial(\triangle \widetilde{f}_{i_1,i_2})}{\partial \lambda_r} &= \frac{\partial}{\partial \lambda_r} \left(- \frac{\widetilde{\alpha}_{i_1,i_2} + \lambda_r \widetilde{\beta}_{i_1,i_2} + \widetilde{b}_{i_1,i_2}}{\widetilde{g}_{0,0} + \lambda_r \widetilde{l}_{0,0}} \right) \\
&= \frac{(\widetilde{\alpha}_{i_1,i_2} + \widetilde{b}_{i_1,i_2}) \widetilde{l}_{0,0} - \widetilde{\beta}_{i_1,i_2} \widetilde{g}_{0,0}}{(\widetilde{g}_{0,0} + \lambda_r \widetilde{l}_{0,0})^2}
\end{aligned}
\tag{5.34}
$$

We can see from equation (5.34) that the evaluation of the derivatives depend on variables which have already been pre-computed for the purpose of the primary restoration. For example, the weights $\widetilde{g}_{0,0}$, $\widetilde{l}_{0,0}$ and the bias $\widetilde{b}_{i_1,i_2}$ are pre-computed at the beginning of the restoration process, and the quantities $\widetilde{\alpha}_{i_1,i_2}$ and $\widetilde{\beta}_{i_1,i_2}$ are already evaluated for the purpose of determining the instantaneous pixel value change $\triangle \widetilde{f}_{i_1,i_2}$ in equation (5.23). As a result, the adaptive regularization does not involve excessive computational overhead.

The size of the training set $\mathcal{V}_r$ depends on the extent of the region $\mathcal{R}_r$ in definition (5.27). It is possible to define $\mathcal{R}_r$ to include the entire image lattice, which would amount to a fixed regularization scheme where we will search for the optimum global regularization parameter value using a single model-based sub-network, but for the problem of adaptive regularization, the region $\mathcal{R}_r$ is usually restricted to a subset of the image lattice, and several such regions are defined for the entire lattice to form an image partition. We associate each such region with a sub-network, the totality of which forms a model-based neural network with hierarchical architecture. In general, we would expect the emergence of regional λ_r values which results in the improved visual quality for the associated region through the training process. This in turn depends critically on the definition of our desired output $\triangle \widetilde{f}^d_{i_1,i_2}$ in the regional training set and our image partition. These two issues will be addressed in Sections 5.7 and 5.8.

5.7   Regional Training Set Definition

To complete the definition of the regional training set $\mathcal{V}_r$, we should supply the desired output $\triangle \widetilde{f}^d_{i_1,i_2}$ for each input vector $\widetilde{\mathbf{f}}_{i_1,i_2}$ in the set. The exact value of $\triangle \widetilde{f}^d_{i_1,i_2}$ which would lead to an optimal visual quality for the particular region concerned is normally unknown due to the usually unsupervised nature of image restoration problems. Nevertheless, an appropriate approximation of this value can usually be obtained by employing a *neighborhood-based prediction scheme* to estimate $\triangle \widetilde{f}^d_{i_1,i_2}$ for the current pixel. This is in the same spirit as the neighborhood-based estimation technique widely used in non-linear filtering applications where a non-linear function defined on a specified neighborhood of the current pixel is used to recover the correct gray level value from its noise-corrupted value [104]. The non-linear filters are usually designed with the purpose of noise suppression in mind. The resulting operations thus have a tendency to over-smooth the edge and textured regions of the image. Remedies to this over-smoothing problem include various edge adaptive filtering schemes where the prediction is performed along the current edge orientation to avoid filtering across the edges [19, 104]. In this work, we would similarly adopt two different prediction schemes for the smooth and textured regions and use the resulting estimated values as the desired outputs for the respective regional training sets.

For combined edge/textured regions, we shall adopt a prediction scheme which emphasizes the dynamic range of the associated region, or equivalently, a scheme which biases toward large values of $\triangle \widetilde{f}^d_{i_1,i_2}$, but at the same time suppresses excessive noise occurring in those regions. For the edge/texture prediction scheme, the following *prediction neighborhood set* for each pixel was adopted.

$$\mathcal{N}_p = \{(k,l) : k,l = -L_p, \ldots, 0, \ldots, L_p\} \tag{5.35}$$

In addition, we define the following mean gray level value $\overline{\widetilde{f}}_{i_1,i_2}$ with respect to this neighborhood set as follows:

$$\overline{\widetilde{f}}_{i_1,i_2} = \frac{1}{N_p} \sum_{k=-L_p}^{L_p} \sum_{l=-L_p}^{L_p} \widetilde{f}_{i_1+k,i_2+l} \tag{5.36}$$

where $\widetilde{f}_{i_1,i_2}$ denotes the gray level value at (i_1,i_2) and $N_p = (2L_p + 1)^2$.

To avoid the problem of over-smoothing in the combined edge/textured regions while suppressing excessive noise at the same time, we have adopted the concept of *weighted order statistic (WOS) filter* [129, 130] in deriving a suitable desired network output $\triangle \widetilde{f}^d_{i_1,i_2}$ for the training set. The set of order statistics corresponding to the prediction neighborhood set can be defined as follows: for the n_p-th order statistic $\widetilde{f}_{i_1,i_2}^{(n_p)}$

$$n_p = P(k,l) \quad \exists (k,l) \in \mathcal{N}_p \tag{5.37}$$

where $P : \mathcal{N}_p \longrightarrow \{1,\ldots,N_p\}$ is a one-to-one mapping such that the following condition is satisfied:

$$\widetilde{f}_{i_1,i_2}^{(1)} \leq \cdots \leq \widetilde{f}_{i_1,i_2}^{(n_p)} \leq \cdots \leq \widetilde{f}_{i_1,i_2}^{(N_p)} \tag{5.38}$$

The output of a weighted order statistic filter is defined as the linear combination of the order statistics

$$\widetilde{f}_{i_1,i_2}^{d} = \sum_{n_p=1}^{N_p} \omega^{(n_p)} \widetilde{f}_{i_1,i_2}^{(n_p)} \tag{5.39}$$

For odd N_p, the simplest example of a WOS filter is the median filter [104] where

$$\omega^{(M_p)} = 1$$
$$\omega^{(n_p)} = 0 \quad n_p \neq M_p \tag{5.40}$$

and

$$M_p = \frac{N_p + 1}{2} \tag{5.41}$$

Therefore, the WOS filter can be considered as a generalization of the median filter where the information from all the order statistics are combined to provide an improved estimate of a variable. In general, for the purpose of noise filtering, the filter weights $\omega^{(1)}$ and $\omega^{(N_p)}$ are chosen such that $\omega^{(1)} \approx 0$ and $\omega^{(N_p)} \approx 0$, since the corresponding order statistics $\widetilde{f}_{i_1,i_2}^{(1)}$ and $\widetilde{f}_{i_1,i_2}^{(N_p)}$ usually represent outliers.

Adopting the value $L_p = 1$ and $N_p = (2 \cdot 1 + 1)^2 = 9$ as the size of the prediction neighborhood set, we define the *predicted gray level value* $\widetilde{f}_{i_1,i_2}^{d}$ and the corresponding desired network output $\triangle \widetilde{f}_{i_1,i_2}^{d}$ according to the operation of a WOS filter as follows:

$$\widetilde{f}_{i_1,i_2}^{d} = \begin{cases} \widetilde{f}_{i_1,i_2}^{(3)} & \widetilde{f}_{i_1,i_2} < \overline{\widetilde{f}}_{i_1,i_2} \\ \widetilde{f}_{i_1,i_2}^{(7)} & \widetilde{f}_{i_1,i_2} \geq \overline{\widetilde{f}}_{i_1,i_2} \end{cases} \tag{5.42}$$

$$\triangle \widetilde{f}_{i_1,i_2}^{d} = \widetilde{f}_{i_1,i_2}^{d} - \widetilde{f}_{i_1,i_2} \tag{5.43}$$

or equivalently

$$\widetilde{f}_{i_1,i_2}^{d} = \sum_{n_p=1}^{9} \omega^{(n_p)} \widetilde{f}_{i_1,i_2}^{(n_p)} \tag{5.44}$$

where

$$\omega^{(3)} = 1, \omega^{(n_p)} = 0, n_p \neq 3 \quad \text{for } \widetilde{f}_{i_1,i_2} < \overline{\widetilde{f}}_{i_1,i_2}$$
$$\omega^{(7)} = 1, \omega^{(n_p)} = 0, n_p \neq 7 \quad \text{for } \widetilde{f}_{i_1,i_2} \geq \overline{\widetilde{f}}_{i_1,i_2}$$

The motivation for choosing the respective order statistics for the two different cases is that, for the case $\widetilde{f}_{i_1,i_2} \geq \overline{\widetilde{f}}_{i_1,i_2}$, we assume that the true gray level value lies in the interval $[\overline{\widetilde{f}}_{i_1,i_2}, \widetilde{f}^{(9)}_{i_1,i_2}]$. For blurred or partially restored image, this corresponds approximately to the interval $[\widetilde{f}^{(5)}_{i_1,i_2}, \widetilde{f}^{(9)}_{i_1,i_2}]$ with its endpoints at the median and maximum gray level values.

Within this interval, we cannot choose $\widetilde{f}^d_{i_1,i_2} = \widetilde{f}^{(5)}_{i_1,i_2} \approx \overline{\widetilde{f}}_{i_1,i_2}$, as this will result in excessive smoothing for the combined edge/textured regions. Neither can we choose $\widetilde{f}^d_{i_1,i_2} = \widetilde{f}^{(9)}_{i_1,i_2}$ with the corresponding order statistic being usually considered an outlier which does not accurately reflect the correct gray level value of the current pixel. A possible candidate could be $\widetilde{f}^d_{i_1,i_2} = 0.5(\widetilde{f}^{(5)}_{i_1,i_2} + \widetilde{f}^{(9)}_{i_1,i_2})$, but the presence of the outlier $\widetilde{f}^{(9)}_{i_1,i_2}$ in the linear combination will still result in non-representative predictions, especially for high levels of noise. To ensure the comparative noise immunity of the resulting estimate while avoiding excessive smoothing, the choice $\widetilde{f}^d_{i_1,i_2} = \widetilde{f}^{(7)}_{i_1,i_2}$ represents a compromise which offers the additional advantage that the gray level value is among one of those in the prediction neighborhood set. The adoption of this value thus implicitly imposes a continuity constraint between the current and the neighboring pixels. The choice of $\widetilde{f}^d_{i_1,i_2} = \widetilde{f}^{(3)}_{i_1,i_2}$ for the case $\widetilde{f}_{i_1,i_2} < \overline{\widetilde{f}}_{i_1,i_2}$ is similarly justified.

On the other hand, we should adopt a prediction scheme which biases toward small values of $\triangle \widetilde{f}^d_{i_1,i_2}$ for the smooth regions to suppress the more visible noise and ringing there. In view of this, the following prediction scheme is adopted for the smooth regions

$$\widetilde{f}^d_{i_1,i_2} = \overline{\widetilde{f}}_{i_1,i_2} \tag{5.45}$$

$$\triangle \widetilde{f}^d_{i_1,i_2} = \widetilde{f}^d_{i_1,i_2} - \overline{\widetilde{f}}_{i_1,i_2} \tag{5.46}$$

This prediction scheme essentially employs the local mean, which serves as a useful indicator of the correct gray level values in smooth regions, as an estimate for the current gray level value. Alternatively, it can be viewed as the operation of a filter mask with all its coefficients being $\frac{1}{N_p}$.

The essential difference between the current approach and traditional adaptive non-linear filtering techniques [104] is that, whereas the traditional filtering techniques *replace* the current gray level value with the above predicted value, the current scheme use this predicted value as a training *guidance* by incorporating it as the desired output in the regional training set. We then apply steepest descent to change the corresponding regional regularization parameter according to the information in these training patterns. In this way, both the information of the degradation mechanism (in the form of the model-based neuronal weights $\widetilde{p}_{k,l}$) and the regional image model (in the form of the regional training set $\mathcal{V}_r$) are exploited to achieve a more accurate restoration.

5.8   Determination of the Image Partition

Conforming to the description of an image as a combination of edge/textured and smooth components, we denote the edge/textured components by $\mathcal{F}_{r_f}, r_f = 1, \ldots, R_f$ and the smooth components by $\mathcal{B}_{r_b}, r_b = 1, \ldots, R_b$. We further define the following combinations of these components

$$\mathcal{F} = \bigcup_{r_f} \mathcal{F}_{r_f} \tag{5.47}$$

$$\mathcal{B} = \bigcup_{r_b} \mathcal{B}_{r_b} \tag{5.48}$$

$$\mathcal{P}_F = \{\mathcal{F}_{r_f}, r_f = 1, \dots, R_f\} \tag{5.49}$$

$$\mathcal{P}_B = \{\mathcal{B}_{r_b}, r_b = 1, \dots, R_b\} \tag{5.50}$$

$$\mathcal{P}_R = \mathcal{P}_F \cup \mathcal{P}_B \tag{5.51}$$

Our partitioning strategy is to first classify each pixel as belonging to the region $\mathcal{F}$ or $\mathcal{B}$ and then derive the partitions $\mathcal{P}_F$ and $\mathcal{P}_B$ using connectivity criteria. We perform the preliminary classification by adopting the following local activity measure δ_{i_1,i_2} for each pixel

$$\delta_{i_1,i_2} = \ln(\sigma_{i_1,i_2}) \tag{5.52}$$

where

$$\sigma_{i_1,i_2} = \left(\frac{1}{N_p} \sum_{k=-L_p}^{L_p} \sum_{l=-L_p}^{L_p} (\widetilde{f}_{i_1+k,i_2+l} - \overline{\widetilde{f}}_{i_1,i_2})^2 \right)^{\frac{1}{2}} \tag{5.53}$$

and $\overline{\widetilde{f}}_{i_1,i_2}$ is defined in equation (5.36). The logarithm mapping is adopted to approximate the non-linear operation in the human vision system which transforms intensity values to perceived contrasts [2]. We then assign the current pixel to either $\mathcal{F}$ or $\mathcal{B}$ according to the value of δ_{i_1,i_2} relative to a threshold T

$$\mathcal{F} = \{(i_1, i_2) : \delta_{i_1,i_2} > T\} \tag{5.54}$$

$$\mathcal{B} = \{(i_1, i_2) : \delta_{i_1,i_2} \leq T\} \tag{5.55}$$

The threshold T is usually selected according to some optimality criteria based on the particular image content. In this work, we have chosen $T = T^*$, where T^* is the threshold which minimizes the total within-class variance ϱ^2 of δ_{i_1,i_2}, i.e.,

$$T^* = \arg\min_T \varrho^2(T) \tag{5.56}$$

where

$$\varrho^2 = \frac{1}{|\mathcal{B}|} \sum_{(i_1,i_2) \in \mathcal{B}} (\sigma_{i_1,i_2} - \sigma_{\mathcal{B}})^2 + \frac{1}{|\mathcal{F}|} \sum_{(i_1,i_2) \in \mathcal{F}} (\sigma_{i_1,i_2} - \sigma_{\mathcal{F}})^2 \tag{5.57}$$

and

$$\sigma_{\mathcal{B}} = \frac{1}{|\mathcal{B}|} \sum_{(i_1,i_2) \in \mathcal{B}} \sigma_{i_1,i_2} \tag{5.58}$$

$$\sigma_{\mathcal{F}} = \frac{1}{|\mathcal{F}|} \sum_{(i_1,i_2) \in \mathcal{F}} \sigma_{i_1,i_2} \tag{5.59}$$

The partitions $\mathcal{P}_F$ and $\mathcal{P}_B$ are in turn extracted from the sets $\mathcal{F}$ and $\mathcal{B}$ by considering each element of $\mathcal{P}_F$ or $\mathcal{P}_B$ to be a maximally connected component of $\mathcal{F}$ or $\mathcal{B}$. To be more precise, if we adopt the usual 8-path connectivity relation $\mathcal{C}_8$ as our connectivity criterion in the case of $\mathcal{F}$ [4], then the partition $\mathcal{P}_F$ will be given by the *quotient set* $\mathcal{F}/\mathcal{C}_8$ of $\mathcal{F}$ under the equivalence relation $\mathcal{C}_8$. The partition $\mathcal{P}_B$ is similarly defined.

In view of the simplicity of this preliminary segmentation scheme and the fact that the relative proportion of the two region types would vary during the course of image restoration, we should adopt some re-segmentation procedures throughout the restoration process to account for this variation. In this work, we have used a modified version of the *nearest neighbor classification procedure* [131] to perform the re-assignment, and we have restricted the re-assignment to pixels on the region boundary. The following *neighboring region set* $\mathcal{N}^{\mathcal{R}}_{i_{b1},i_{b2}} \subset \mathcal{P}_R$ is defined for each such boundary pixel

$$\mathcal{N}^{\mathcal{R}}_{i_{b1},i_{b2}} = \{\mathcal{R}_q, q = 1, \ldots, Q\} \qquad (5.60)$$

where each region $\mathcal{R}_q$ in the set is adjacent to the boundary pixel (i_{b1}, i_{b2}), or more precisely, there exists at least one pixel in each $\mathcal{R}_q$ which is in the 8-neighborhood of (i_{b1}, i_{b2}). Corresponding to each $\mathcal{R}_q$ we can define the following regional activity $\overline{\sigma}_q$

$$\overline{\sigma}_q = \frac{1}{|\mathcal{R}_q|} \sum_{(i_1,i_2)\in\mathcal{R}_q} \sigma_{i_1,i_2} \qquad (5.61)$$

where σ_{i_1,i_2} is defined in equation (5.53). With the definition of these variables, we can proceed with the re-classification of the boundary pixels by adopting the following nearest neighbor decision rule for (i_{b1}, i_{b2})

$$(i_{b1}, i_{b2}) \in \mathcal{R}_{q^*} \quad \text{if } |\sigma_{i_{b1},i_{b2}} - \overline{\sigma}_{q^*}| < |\sigma_{i_{b1},i_{b2}} - \overline{\sigma}_q|, q = 1, \ldots, Q \qquad (5.62)$$

The application of this decision rule is manifested as a continual change in the boundary of each region in such a way as to increase the homogeneity of the activities of the regions. Finally, with the full determination of the various texture and smooth components in the image by the above scheme, we can apply the steepest descent rule (5.29) to each regional regularization parameter λ_r and using the prediction schemes for the respective regions to achieve adaptive regularization.

5.9   The Edge-Texture Characterization Measure

In this section, we introduce the Edge-Texture Characterization (ETC) measure, which is a scalar quantity summarizing the degree of resemblance of a particular pixel value configuration to either textures or edges. In other words, pixel value arrangements corresponding to textures and edges will in general exhibit different values for this measure. This is unlike the case where the local variance or the edge magnitude [2] is adopted for the image activity measure. Due to the possibility that both edges and textures may exhibit similar levels of image activities in terms of gray level variations around their neighborhoods, it is usually not possible to distinguish between these two feature types using the conventional image activity measures.

On the other hand, the current ETC measure is derived based on the correlational properties of individual pixel value configurations. In general, we may expect that configurations corresponding to edges and textures will possess significantly different correlational properties, with the individual pixels in a texture configuration being far less correlated with each other than those in an edge configuration. This is described in a quantitative way using the current measure. More importantly, we have analytically established intervals of ETC measure values corresponding to those pixel configurations which visually more resemble textures than edges, and *vice versa*.

This measure is especially useful for distinguishing between edges and textures in image restoration such that we can specify different levels of regularization to each of them. Due to the different noise masking capabilities of these two feature types, it is usually not desirable to apply similar values of regularization parameters to both of them. With the incorporation of the newly formulated ETC measure into our HMBNN approach, we are able to separately estimate two different parameter values which are optimal to edges and textures, respectively, and apply the correct parameter to the current pixel in accordance with its associated ETC measure value.

The starting point of our formulation is as follows: we consider the gray level values of image pixels in a local region as i.i.d. random variables with variance σ^2. If we apply a local $K \times K$ averaging operation to each pixel, the variance σ'^2 of the smoothed random variables is given by

$$\sigma'^2 = \frac{1}{K^4} \mathbf{u}^T \mathbf{R} \mathbf{u} = \frac{\sigma^2}{K^2} \tag{5.63}$$

where $\mathbf{R} = diag[\sigma^2, \ldots, \sigma^2]$ is the $K^2 \times K^2$ covariance matrix of the K^2 random variables in the $K \times K$ averaging window, and $\mathbf{u}$ is an $K^2 \times 1$ vector with all entries equal to one. The diagonal structure of the covariance matrix is due to the independence of the random variables. The i.i.d. assumption above is in general not applicable to real-world images. In fact, we usually identify a meaningful image with the existence of controlled correlation among its pixels. As a result, we generalize the above i.i.d. case to incorporate correlations inside the $K \times K$ window.

We define the multiset $\mathcal{P} = \{P_1, \ldots, P_i, \ldots, P_m\}$, where $P_1 + \ldots + P_m = K^2$, as a *partition* of the K^2 variables in the window into m components. In addition, we assume that all variables within the ith component is correlated with each other with correlation coefficient ρ_i, and variables among different components are mutually uncorrelated with each other. Some examples of $\mathcal{P}$ in a 5×5 window are given in Figure 5.5. For example, we can describe the region around an edge pixel by the partition $\mathcal{P} = \{P_1, P_2\}$, where $P_1 \approx P_2$ (Figure 5.5a). In

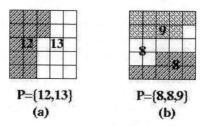

P={12,13}   P={8,8,9}
(a)   (b)

Figure 5.5: Illustrating the various forms of partition $\mathcal{P}$

this general case, the variance σ'^2 after $K \times K$ averaging is given by

$$\sigma'^2 = \frac{1}{K^4} \mathbf{u}^T R \mathbf{u} = \frac{\sigma^2}{\kappa^2} \tag{5.64}$$

In this case, $\mathbf{R}$ is a *block-diagonal* matrix with the following structure

$$\mathbf{R} = diag[\mathbf{R}_1, \dots, \mathbf{R}_m] \tag{5.65}$$

Each sub-matrix $\mathbf{R}_i, i = 1, \dots, m$ is of dimension $P_i \times P_i$ with the following structure

$$\mathbf{R}_i = \sigma^2 \begin{bmatrix} 1 & \rho_i & \cdots & \rho_i \\ \rho_i & \ddots & \ddots & \vdots \\ \vdots & \ddots & & \rho_i \\ \rho_i & \cdots & \rho_i & 1 \end{bmatrix} \tag{5.66}$$

If we carry out the matrix multiplication in equation (5.64), the square of the quantity κ in the equation is evaluated to be

$$\kappa^2 = \frac{K^4}{K^2 + \sum_{P_i \in \mathcal{P}} \rho_i (P_i^2 - P_i)} \tag{5.67}$$

Assuming that $0 \leq \rho_i \leq 1$ for all i, which implies positive correlation among pixels within a single component, the value of κ is maximized when $\rho_i = 0, \forall i$, giving $\kappa = K$ in equation (5.67), which corresponds to the previous case of i.i.d. variables.

On the other hand, if we assume $\rho_i = 1$ for all i within a single element partition $\mathcal{P} = \{K^2\}$ and substituting into equation (5.67), we have

$$\kappa^2 = \frac{K^4}{K^2 + (K^4 - K^2)} = 1 \tag{5.68}$$

which implies $\kappa = 1$. This corresponds to the case where all the gray level values within the window are highly correlated, or in other words, to smooth regions in the image. In general, the value of κ is between 1 and K. Thus it serves as an indicator of the degree of correlation within the $K \times K$ window. Larger values of κ indicate low level of correlation among the pixels in the window, which are usually the cases for textured and edge regions, while smaller values of κ usually correspond to smooth regions as indicated above. To provide an intuitive grasp of the values of κ corresponding to various features in the image, we carry out the calculation prescribed in equation (5.67) for a 5×5 averaging window.

For $K = 5$, the minimum and maximum values of κ are 1 and 5, respectively. For a positive within-component correlation coefficient ρ_i, the value of κ is constrained within the interval $[1, 5]$. Referring to Figure 5.5a which describes image edge regions with the partition $\mathcal{P} = \{12, 13\}$, and further assumes that $\rho_i = \rho = 0.9$ for all components, we have, after substituting the corresponding quantities into equation (5.67)

$$\kappa^2 = \frac{5^4}{5^2 + 0.9[(12^2 - 12) + (13^2 - 13)]} \approx 2.20 \tag{5.69}$$

or $\kappa \approx 1.48$. This value of κ, which we designate as κ_2, serves to characterize all edge-like features in the image if a 5×5 averaging window is used. On the other hand, if we consider more complex features with the number of components $m > 2$, which usually correspond to textures in the images, we should expect the value of κ to be within the interval $[1.48, 5]$. This is confirmed by evaluating κ for the partition $\mathcal{P} = \{8, 8, 9\}$ as illustrated in Figure 5.5b, again assuming $\rho_i = 0.9$ for all i,

$$\kappa^2 = \frac{5^4}{5^2 + 0.9[2(8^2 - 8) + (9^2 - 9)]} \approx 3.28 \tag{5.70}$$

or $\kappa \approx 1.81$, which we designate as κ_3. As a result, the value of κ indicates to a certain extent the qualitative attributes of a local image region, i.e., whether it is more likely to be a textured region or an edge region, rather than just distinguishing the smooth background from the combined texture/edge regions as in the case of using the local standard deviation σ_{i_1, i_2} alone. We can therefore refer to this quantity κ as the Edge-Texture Characterization (ETC) measure.

Table 5.1 lists the values of the measure κ_1 to κ_9 and κ_{25}. We can notice that, in general, the value of κ increases with the number of correlated components within the pixel window, which confirms our previous observation. The case of 25 components corresponds to the case of i.i.d. random variables and results in the value $\kappa_{25} = 5$. It is also noted that the value of κ_1 is slightly larger than the

Table 5.1: ETC measure values for various different partitions

κ_1	1.05	κ_6	2.54
κ_2	1.48	κ_7	2.72
κ_3	1.81	κ_8	2.91
κ_4	2.08	κ_9	3.07
κ_5	2.33	κ_{25}	5.00

case of fully correlated random variables in our previous calculation due to our present assumption of $\rho = 0.9$.

We can estimate the value of κ in a pixel neighborhood by the ratio $\hat{\sigma}/\hat{\sigma}'$ in accordance with equation (5.64), where $\hat{\sigma}$ and $\hat{\sigma}'$ are the sample estimates of σ and σ', respectively.

$$\hat{\sigma}^2 = \frac{1}{|\mathcal{N}|} \sum_{(i,j)\in\mathcal{N}} (\widetilde{f}_{i,j} - \overline{\widetilde{f}})^2 \tag{5.71}$$

$$\hat{\sigma}'^2 = \frac{1}{|\mathcal{N}|} \sum_{(i,j)\in\mathcal{N}} (\widetilde{f}'_{i,j} - \overline{\widetilde{f}'})^2 \tag{5.72}$$

In the equations, $\mathcal{N}$ denotes a neighborhood set around the current pixel, $\widetilde{f}_{i,j}$ denotes the gray level value of pixel (i,j) in the set, and $\widetilde{f}'_{i,j}$ is the corresponding smoothed gray level value under $K \times K$ averaging. $\overline{\widetilde{f}}$ and $\overline{\widetilde{f}'}$ are, respectively, the mean of the gray level values of the original and smoothed variables in the neighborhood set. In general, this empirically estimated κ is not restricted to the interval $[1, K]$ due to the use of the sample variances, but most of its values are restricted to the interval $[0, K]$.

5.10   The ETC Fuzzy HMBNN for Adaptive Regularization

In this section, we propose a fuzzified version of the model-based neural network for adaptive regularization. A hierarchical architecture is again adopted for the current network, but we have included two neurons in each sub-network instead of a single neuron. We referred to those two neurons as the edge neuron and the texture neuron, respectively. They in turn estimate *two* regularization parameters, namely, the edge parameter and the texture parameter, for each region with values optimal for regularizing the edge pixels and the textured pixels, respectively, within the region. Ideally, the parameter of the edge neuron should be updated using the information of the edge pixels only, and the texture neuron should be updated using the texture pixels only, which implies the necessity to distinguish between the edge and texture pixels. This is precisely the motivation for the formulation of the ETC measure, which achieves this very purpose. As the concepts of edge and texture are inherently fuzzy, we characterize this fact by defining two fuzzy sets, the EDGE fuzzy set and the TEXTURE fuzzy set, over the ETC measure domain. This is possible since the value of the ETC measure reflects the degree of resemblance of a particular local gray level configuration to either textures or edges. More importantly, there exists definite intervals of measure values which we can claim that the corresponding gray level configuration should be more like textures, and *vice versa*. As a result, the ETC measure domain serves as an ideal *universe of discourse* [37] for the definition of the EDGE and TEXTURE fuzzy sets. In view of the importance of the fuzzy set concept in the current formulation, we will first have a review on this topic.

5.11 Theory of Fuzzy Sets

The inclusion of a member a in a set A is usually represented symbolically as $a \in A$. Alternatively, we can express this inclusion in terms of a *membership function* $\mu_A(a)$ as follows

$$\mu_A(a) = \begin{cases} 1 & \text{if } a \in A \\ 0 & \text{if } a \notin A \end{cases} \tag{5.73}$$

The membership function takes values in the discrete set $\{0, 1\}$. Zadeh, in his 1965 paper [38], generalized the definition of the membership function such that it can take values in the real-valued interval $[0, 1]$. The generalized set corresponding to the new membership function is known as a *fuzzy set* [36, 37, 38, 132, 133, 134] in contrast with the previous *crisp set*.

The implication of the new membership function is that, aside from the states of belonging wholly to a set or not belonging to a set at all, we can now allow an element to be a member of a fuzzy set "to a certain extent," with the exact degree of membership being expressed by the value of the corresponding membership function. Therefore, if the membership function value $\mu_A(a)$ of an element a is close to 1, we can say that a belongs "to a large degree" to the fuzzy set A, and we should expect that a possesses many of the properties that are characteristic of the set A. On the other hand, if the membership function value is small, we can expect that the element a only bears a vague resemblance to a typical member of the set A.

This concept is particularly important for expressing how human beings characterize everyday concepts. Take for example the concept "tall." We can immediately visualize the meaning of the word. But in order for machine interpretation systems to recognize this concept, we must provide an exact definition. If restricted to the use of crisp sets, we may define this concept in terms of the set $T = \{h : h \geq 1.7m\}$ where h is the height variable. According to this definition, a value of the variable h such as $h = 1.699m$ will not be considered as belonging to the concept "tall," which is certainly unnatural from the human viewpoint. If we instead define "tall" to be a fuzzy set with the membership function shown in Figure 5.6, the value $h = 1.699m$ can still be interpreted as belonging "strongly" to the set, and thus conforms more closely to human interpretation.

The ordinary operations on crisp set can be generalized to the case of fuzzy set in terms of the membership function values as follows:

$$\mu_{A \cap B}(x) = \min\{\mu_A(x), \mu_B(x)\} \tag{5.74}$$

$$\mu_{A \cup B}(x) = \max\{\mu_A(x), \mu_B(x)\} \tag{5.75}$$

$$\mu_{A^c}(x) = 1 - \mu_A(x) \tag{5.76}$$

where $A \cap B$, $A \cup B$ and A^c are, respectively, the intersection of fuzzy sets A and B the union of A and B and the complement of A. These equations reduce to the ordinary intersection, union and complement operations on crisp sets when the ranges of the membership functions $\mu_A(x), \mu_B(x)$ are restricted to values in $\{0, 1\}$.

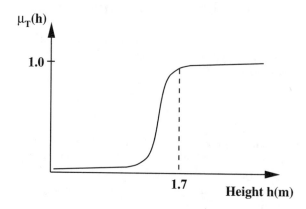

Figure 5.6: The fuzzy set representing the concept TALL

A *fuzzy inference relationship* is usually of the following form:

> If $((x_1$ has property $A_1)$ ⊗ ...... ⊗ $(x_n$ has property $A_n))$, then $(y$ has property $B)$

where $x_1, \ldots, x_n$ and y are numerical variables, $A_1, \ldots, A_n$ and B are *linguistic* descriptions of the properties required of the corresponding numerical variables, and ⊗ denotes either the union or intersection operations. As described above, we can convert the linguistic descriptions $A_1, \ldots, A_n$ and B into fuzzy sets with membership functions $\mu_{A_1}(x_1), \ldots, \mu_{A_n}(x_n)$ and $\mu_B(y)$, where we have identified the names of the fuzzy sets with their corresponding linguistic descriptions. Formally, we can describe this inference operation as a mapping between fuzzy sets as follows:

$$B' = F(A_1, \ldots, A_n) \tag{5.77}$$

where $B' \subset B$ is a *fuzzy subset* of B. Fuzzy subsethood is formally described by the following condition

$$A \subset B \text{ if } \mu_A(x) \le \mu_B(x), \forall x \tag{5.78}$$

which reduces to the ordinary subsethood relationship between crisp sets if $\mu_A(x), \mu_B(x)$ are allowed to take values only in $\{0, 1\}$.

The particular fuzzy subset B' chosen within the fuzzy set B (or equivalently, the particular form of the mapping F) depends on the degree to which the current value of each variable $x_i, i = 1, \ldots, n$ belongs to their respective fuzzy sets A_i. To summarize, the inference procedure accepts fuzzy sets as inputs and emits a single fuzzy set as output.

In practical systems, a *numerical* output which captures the essential characteristics of the output fuzzy set B' is usually required. This is usually done by specifying a *defuzzification* operation D on the fuzzy set B' to produce the numerical output y'

$$y' = D(B') \tag{5.79}$$

A common defuzzification operation is the following *centroid defuzzification* [37] operation

$$y' = \frac{\int_{-\infty}^{\infty} y \mu_{B'}(y) \, dy}{\int_{-\infty}^{\infty} \mu_{B'}(y) \, dy} \tag{5.80}$$

where we assign the centroid of the fuzzy membership function $\mu_{B'}(y)$ to the variable y'.

5.12 Edge-Texture Fuzzy Model Based on ETC Measure

We have previously defined the Edge-Texture Characterization (ETC) measure κ which quantifies the degree of resemblance of a particular gray level configuration to either textures or edges. In addition, we have established that, for values of κ within the interval $I_2 = (\frac{\kappa_1 + \kappa_2}{2}, \kappa_3]$, we can consider the underlying gray level configuration to be reasonably close to that of edges, and for $\kappa > \kappa_3$, we can conclude that the corresponding configuration has a closer resemblance to that of textures. However, if we consider the value $\kappa = \kappa_3 + \epsilon$, where ϵ is a small positive constant, we will classify the corresponding configuration as a texture configuration, but we can expect that it will still share many of the properties of an edge configuration due to the closeness of κ to an admissible edge value. In fact, it is difficult to define the concepts of "edge" and "textures" in terms of crisp sets in the ETC domain. In view of this, fuzzy set theory becomes a natural candidate for characterizing these concepts in terms of the ETC measure.

We therefore define two fuzzy sets, namely, the EDGE fuzzy set and the TEXTURE fuzzy set, on the ETC measure domain in terms of their membership functions $\mu_E(\kappa)$ and $\mu_T(\kappa)$, as follows:

$$\mu_E(\kappa) \equiv \frac{1}{1 + e^{\beta_E(\kappa - \kappa_E)}} \tag{5.81}$$

$$\mu_T(\kappa) \equiv \frac{1}{1 + e^{-\beta_T(\kappa - \kappa_T)}} \tag{5.82}$$

The two set membership functions are plotted in Figure 5.7. From the figure, it is seen that $\mu_E(\kappa)$ is a decreasing sigmoid function with the transition point at $\kappa = \kappa_E$, and $\mu_T(\kappa)$ is an increasing sigmoid function with the transition point at $\kappa = \kappa_T$. The parameters β_E and β_T control the steepness of transition of the two membership functions. In view of our previous observation which characterizes edges with values of κ within the interval I_2, we may have expected the function $\mu_E(\kappa)$ to exhibit a peak around κ_2 and to taper off on both sides. Instead, we have chosen a decreasing sigmoid function with the transition point at $\kappa_E \approx \kappa_2$, which implicitly classifies the smooth regions with $\kappa < \kappa_2$ as belonging to the EDGE fuzzy set as well. This is due to our formulation of the current regularization algorithm in such a way that, whenever a certain pixel configuration has a larger membership value in the EDGE fuzzy set, larger values of regularization parameters will be applied to this configuration. As a result, it is reasonable

Figure 5.7: The EDGE and TEXTURE fuzzy membership functions

to assign greater membership values to those configurations with $\kappa < \kappa_2$, which usually corresponds to weak edges or smooth regions, due to their less effective noise masking capabilities compared with strong edges. In other words, we are interpreting the membership function $\mu_E(\kappa)$ as the indicator of the required amount of regularization for a certain pixel configuration rather than its true correspondence to a characteristic edge configuration.

On the other hand, the shape of the TEXTURE set membership function truly reflects the fact that we consider those gray level configurations with $\kappa > \kappa_3$ to be essentially textures. However, instead of choosing $\kappa_T = \kappa_3$, which was shown to correspond to a gray level configuration containing three uncorrelated components and may reasonably resemble textures, we instead choose the more conservative value of $\kappa_T = 2 > \kappa_3$.

From the two membership functions, we define the following *normalized ETC fuzzy coefficients* $\widetilde{\mu}_E(\kappa)$ and $\widetilde{\mu}_T(\kappa)$

$$\widetilde{\mu}_E(\kappa) = \frac{\mu_E(\kappa)}{\mu(\kappa)} \tag{5.83}$$

$$\widetilde{\mu}_T(\kappa) = \frac{\mu_T(\kappa)}{\mu(\kappa)} \tag{5.84}$$

where

$$\mu(\kappa) = \mu_E(\kappa) + \mu_T(\kappa) \tag{5.85}$$

5.13 Architecture of the Fuzzy HMBNN

Corresponding to the partition of the image into the combined edge/texture components $\mathcal{F}_{r_f}, r_f = 1, \ldots, R_f$ and the smooth regions $\mathcal{B}_{r_b}, r_b = 1, \ldots, R_b$, we assign individual sub-networks to each of these regions. For the smooth regions, we assign one neuron to each smooth region to estimate a single parameter $\lambda_{r_b}, r_b = 1, \ldots, R_b$ for each region. However, for the combined edge/textured regions, instead of employing only one neuron as in the previous case, we have assigned *two* neurons, which we designate as the edge neuron and the texture neuron with associated weight vectors $\widetilde{\mathbf{p}}_{edge}(\lambda_{r_f}^{edge})$ and $\widetilde{\mathbf{p}}_{tex}(\lambda_{r_f}^{tex})$, to each edge/texture sub-network. The two weight vectors are, respectively, functions of the *edge regularization parameter* $\lambda_{r_f}^{edge}$ and the *texture regularization parameter* $\lambda_{r_f}^{tex}$. As their names imply, we should design the training procedure in such a way that the parameter $\lambda_{r_f}^{edge}$ estimated through the edge neuron should be optimal for the regularization of edge-like entities, and $\lambda_{r_f}^{tex}$ estimated through the texture neuron should be optimal for textured regions.

Corresponding to these two weight vectors, we evaluate two estimates of the required pixel change, $\triangle \widetilde{f}_{i_1,i_2}^{edge}$ and $\triangle \widetilde{f}_{i_1,i_2}^{tex}$, for the same gray level configuration $\widetilde{\mathbf{f}}_{i_1,i_2}$ as follows

$$\triangle \widetilde{f}_{i_1,i_2}^{edge} = \widetilde{\mathbf{p}}_{edge}(\lambda_{r_f}^{edge})^T \widetilde{\mathbf{f}}_{i_1,i_2} \tag{5.86}$$

$$\triangle \widetilde{f}_{i_1,i_2}^{tex} = \widetilde{\mathbf{p}}_{tex}(\lambda_{r_f}^{tex})^T \widetilde{\mathbf{f}}_{i_1,i_2} \tag{5.87}$$

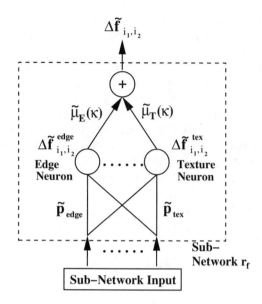

Figure 5.8: The architecture of the fuzzy HMBNN sub-network

where $(i_1, i_2) \in \mathcal{F}_{r_f}$. The quantities $\triangle \widetilde{f}_{i_1,i_2}^{edge}$ and $\triangle \widetilde{f}_{i_1,i_2}^{tex}$ are the required updates based on the assumptions that the underlying gray level configuration $\widetilde{\mathbf{f}}_{i_1,i_2}$ corresponds to edge or textures, respectively.

Referring to the fuzzy formulation in the previous section, we can evaluate the ETC measure κ_{i_1,i_2} at pixel position (i_1, i_2) as a function of the pixel values in a local neighborhood of (i_1, i_2). From this we can calculate the normalized edge fuzzy coefficient value $\widetilde{\mu}_E(\kappa_{i_1,i_2})$ and the normalized texture fuzzy coefficient value $\widetilde{\mu}_T(\kappa_{i_1,i_2})$. As a result, a natural candidate for the final required pixel update value $\triangle \widetilde{f}_{i_1,i_2}$ can be evaluated as a convex combination of the quantities $\triangle \widetilde{f}_{i_1,i_2}^{edge}$ and $\triangle \widetilde{f}_{i_1,i_2}^{tex}$ with respect to the normalized fuzzy coefficient values

$$\triangle \widetilde{f}_{i_1,i_2} = \widetilde{\mu}_E(\kappa_{i_1,i_2})\triangle \widetilde{f}_{i_1,i_2}^{edge} + \widetilde{\mu}_T(\kappa_{i_1,i_2})\triangle \widetilde{f}_{i_1,i_2}^{tex} \tag{5.88}$$

From the above equation, it is seen that for regions around edges where $\widetilde{\mu}_E(\kappa_{i_1,i_2}) \approx 1$ and $\widetilde{\mu}_T(\kappa_{i_1,i_2}) \approx 0$, the final required gray level update $\triangle \widetilde{f}_{i_1,i_2}$ is approximately equal to $\triangle \widetilde{f}_{i_1,i_2}^{edge}$, which is optimal for the restoration of the edges. On the other hand, in textured regions where $\widetilde{\mu}_T(\kappa_{i_1,i_2}) \approx 1$ and $\widetilde{\mu}_E(\kappa_{i_1,i_2}) \approx 0$, the required update $\triangle \widetilde{f}_{i_1,i_2}$ assumes the value $\triangle \widetilde{f}_{i_1,i_2}^{tex}$ which is optimal for textures, provided that proper training procedures are adopted for both of the neurons. Alternatively, this equation can be interpreted as the operation of a *two-layer* sub-network, where $\triangle \widetilde{f}_{i_1,i_2}$ is the network output, $\widetilde{\mu}_E(\kappa_{i_1,i_2}), \widetilde{\mu}_T(\kappa_{i_1,i_2})$ are output weights and $\triangle \widetilde{f}_{i_1,i_2}^{edge}$ and $\triangle \widetilde{f}_{i_1,i_2}^{tex}$ are the hidden neuron outputs. The architecture of this two-layer sub-network is shown in Figure 5.8.

5.13.1 Correspondence with the General HMBNN Architecture

Similar to our previous model-based neural network formulation, the current fuzzy HMBNN model also adopts the hidden-node hierarchical architecture, which is depicted in Figure 5.3. The architecture of the sub-networks for the regularization of the smooth regions remains the same. However, for the edge/texture regularization sub-networks, we have adopted a new architecture in the form of a two-layer network, with two neurons in the hidden layer as shown in Figure 5.8. Corresponding to the general form for evaluation of the sub-network output, we can describe the operation of the edge/texture regularization sub-network for region $\mathcal{F}_{r_f}$ as follows:

$$\phi(\widetilde{\mathbf{f}}_{i_1,i_2}, \widetilde{\mathbf{p}}_{r_f}) = \sum_{s_{r_f}=1}^{2} c_{s_{r_f}} \psi_{r_f}(\widetilde{\mathbf{f}}_{i_1,i_2}, \widetilde{\mathbf{p}}(\lambda_{r_f}^{s_{r_f}})) \tag{5.89}$$

where

$$\lambda_{r_f}^1 \equiv \lambda_{r_f}^{edge} \tag{5.90}$$

$$\lambda_{r_f}^2 \equiv \lambda_{r_f}^{tex} \tag{5.91}$$

$$\psi_{r_f}(\widetilde{\mathbf{f}}_{i_1,i_2}, \widetilde{\mathbf{p}}(\lambda_{r_f}^{s_{r_f}})) \equiv \widetilde{\mathbf{p}}(\lambda_{r_f}^{s_{r_f}})^T \widetilde{\mathbf{f}}_{i_1,i_2} \tag{5.92}$$

$$c_1 \equiv \widetilde{\mu}_E(\kappa_{i_1,i_2}) \tag{5.93}$$

$$c_2 \equiv \widetilde{\mu}_T(\kappa_{i_1,i_2}) \tag{5.94}$$

and $\widetilde{\mathbf{p}}_{r_f}$ can be considered as a concatenation of $\widetilde{\mathbf{p}}(\lambda_{r_f}^{s_{r_f}}), s_{r_f} = 1, 2$.

5.14 Estimation of the Desired Network Output

We have previously defined the desired network output $\triangle \widetilde{f}_{i_1,i_2}^d$ and the associated predicted gray level value $\widetilde{f}_{i_1,i_2}^d$, where

$$\widetilde{f}_{i_1,i_2}^d = \widetilde{f}_{i_1,i_2} + \triangle \widetilde{f}_{i_1,i_2}^d \tag{5.95}$$

as follows: for smooth regions, we define the variable $\widetilde{f}_{i_1,i_2}^d$ as the mean of the neighboring pixel values

$$\widetilde{f}_{i_1,i_2}^d = \overline{\widetilde{f}}_{i_1,i_2} \tag{5.96}$$

$$\overline{\widetilde{f}}_{i_1,i_2} = \frac{1}{N_p} \sum_{k=-L_p}^{L_p} \sum_{l=-L_p}^{L_p} \widetilde{f}_{i_1+k,i_2+l} \tag{5.97}$$

where $N_p = (2L_p + 1)^2$ is the size of the filter mask for averaging. For combined edge/textured regions, we consider the ranked pixel values in the prediction neighborhood set

$$\mathcal{N}_p = \{\widetilde{f}_{i_1,i_2}^{(1)}, \ldots, \widetilde{f}_{i_1,i_2}^{(n_p)}, \ldots, \widetilde{f}_{i_1,i_2}^{(N_p)}\}$$

where $\widetilde{f}_{i_1,i_2}^{(n_p)}$ is the n_pth order statistic of the pixels in the ordered prediction neighborhood set and corresponds to $\widetilde{f}_{i_1+k,i_2+l}$ for appropriate k and l. The ranked gray level values satisfy the following condition

$$\widetilde{f}_{i_1,i_2}^{(1)} \leq \cdots \leq \widetilde{f}_{i_1,i_2}^{(n_p)} \leq \cdots \leq \widetilde{f}_{i_1,i_2}^{(N_p)}$$

Adopting a prediction neighborhood size of $N_p = 9$, we define the predicted gray level value $\widetilde{f}_{i_1,i_2}^d$ for the combined edge/textured regions as follows:

$$\widetilde{f}_{i_1,i_2}^d = \begin{cases} \widetilde{f}_{i_1,i_2}^{(3)} & \widetilde{f}_{i_1,i_2} < \overline{\widetilde{f}}_{i_1,i_2} \\ \widetilde{f}_{i_1,i_2}^{(7)} & \widetilde{f}_{i_1,i_2} \geq \overline{\widetilde{f}}_{i_1,i_2} \end{cases} \tag{5.98}$$

It was observed experimentally that this prediction scheme results in a noisy appearance for the restored edges at high noise levels due to its smaller noise masking capability compared with textures. In view of this, we apply the following *edge-oriented* estimation scheme at high noise levels

$$\widetilde{f}_{i_1,i_2}^d = \begin{cases} \widetilde{f}_{i_1,i_2}^{(4)} & \widetilde{f}_{i_1,i_2} < \overline{\widetilde{f}}_{i_1,i_2} \\ \widetilde{f}_{i_1,i_2}^{(6)} & \widetilde{f}_{i_1,i_2} \geq \overline{\widetilde{f}}_{i_1,i_2} \end{cases} \tag{5.99}$$

This edge-oriented scheme results in a less noisy appearance for the edges, but the textured areas will appear blurred compared with the previous case. It would therefore be ideal if, at high noise levels, we can apply the texture-oriented estimation scheme to the textured areas only, and apply the edge-oriented estimation scheme to only the edges. This in turn requires the separation of the edges and textured areas, which is usually difficult in terms of conventional measures such as image gradient magnitudes or local variance due to the similar levels of gray level activities for these two types of features. On the other hand, the previous fuzzy formulation in terms of the scalar ETC measure allows this very possibility.

In addition, equations (5.98) and (5.99) predict the desired gray level value $\widetilde{f}_{i_1,i_2}^d$ in terms of a *single* order statistic only. The estimation will usually be more meaningful if we can exploit information provided by the gray level values of *all* the pixels in the prediction neighborhood set $\mathcal{N}_p$. In the next section, we investigate this possibility by extending the previous crisp estimation framework to a fuzzy estimation framework for $\widetilde{f}_{i_1,i_2}^d$.

5.15   Fuzzy Prediction of Desired Gray Level Value

In this section, we define two fuzzy sets, namely, the EDGE GRAY LEVEL ESTIMATOR (EG) fuzzy set and the TEXTURE GRAY LEVEL ESTIMATOR (TG) fuzzy set, over the domain of *gray level values* in an image. This fuzzy formulation is independent of our previous ETC-fuzzy formulation where the EDGE and TEXTURE fuzzy sets are defined over the domain of the *ETC measure*. The purpose of this second fuzzy formulation is to allow the utilization of different prediction strategies for $\tilde{f}^d_{i_1,i_2}$ in the edge and textured regions, respectively, and the evaluation of this predicted gray level value using all the gray level values in the prediction neighborhood set $\mathcal{N}_p$ instead of a single crisp value.

5.15.1 Definition of the Fuzzy Estimator Membership Function

For this gray level estimator fuzzy model, we define the two set membership functions, $\varphi_{EG}(\tilde{f}_{i_1,i_2})$ and $\varphi_{TG}(\tilde{f}_{i_1,i_2})$ in terms of Gaussian functions, as opposed to the use of sigmoid nonlinearity in the first model. Again denoting the gray level value at location (i_1, i_2) as $\tilde{f}_{i_1,i_2}$ and the n_p-th order statistic of the prediction neighborhood set $\mathcal{N}_p$ as $\tilde{f}^{(n_p)}_{i_1,i_2}$, and assuming that $\tilde{f}_{i_1,i_2} \geq \overline{\tilde{f}}_{i_1,i_2}$ without loss of generality, which corresponds to the second condition in equations (5.98) and (5.99), we define the membership functions of the EG fuzzy set and the TG fuzzy set as follows:

$$\varphi_{EG}(\tilde{f}_{i_1,i_2}) = e^{-\xi_{EG}(\tilde{f}_{i_1,i_2}-\tilde{f}^{(6)}_{i_1,i_2})^2} \qquad (5.100)$$

$$\varphi_{TG}(\tilde{f}_{i_1,i_2}) = e^{-\xi_{TG}(\tilde{f}_{i_1,i_2}-\tilde{f}^{(7)}_{i_1,i_2})^2} \qquad (5.101)$$

For the condition $\tilde{f}_{i_1,i_2} < \overline{\tilde{f}}_{i_1,i_2}$, we will replace the centers of the EG and TG membership functions with $\tilde{f}^{(4)}_{i_1,i_2}$ and $\tilde{f}^{(3)}_{i_1,i_2}$, respectively, in accordance with equations (5.98) and (5.99).

The two membership functions are depicted graphically in Figure 5.9. Under the condition $\tilde{f}_{i_1,i_2} \geq \overline{\tilde{f}}_{i_1,i_2}$, where we previously designate $\tilde{f}^{(6)}_{i_1,i_2}$ as the preferred gray level estimator in the edge-oriented estimation scheme, we now generalize this concept by assigning a membership value of 1 for $\tilde{f}^{(6)}_{i_1,i_2}$. However, instead of adopting this value exclusively, we have also assigned non-zero membership function values for gray level values close to $\tilde{f}^{(6)}_{i_1,i_2}$, thus expressing the notion

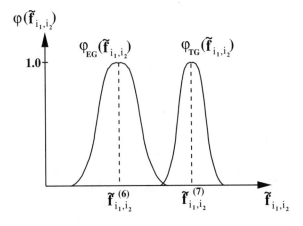

Figure 5.9: The fuzzy membership functions of the sets EG and TG

that nearby values also have relevancy in determining the final predicted gray level value. Similar assignments are adopted for the TG membership function with $\widetilde{f}_{i_1,i_2}^{(7)}$ as the center.

In the two equations, the two parameters ξ_{EG} and ξ_{TG} control the width of the respective membership functions, thus quantifying the notion of "closeness" in terms of gray level value differences for both fuzzy sets. In general, the values of these two parameters are different from each other, and the optimal values of both parameters may also differ for different image regions. In view of these, we have devised a training algorithm for adapting these parameters in different regions and this will be described in a later section. The requirement for adaptation is also the reason we have chosen ξ_{EG} and ξ_{TG} to be multiplicative factors instead of their usual appearances as denominators in the form of variances for Gaussian functions: in the former case, we have only to ensure that both parameters are greater than zero. In the latter case, there is the additional difficulty that the adapting variances can easily become too small to result in an excessively large exponent for the Gaussian functions.

5.15.2   Fuzzy Inference Procedure for Predicted Gray Level Value

Given the membership functions $\varphi_{EG}(\widetilde{f}_{i_1,i_2})$ and $\varphi_{TG}(\widetilde{f}_{i_1,i_2})$ of the two fuzzy sets EG and TG, we can define a mapping F, which assigns a single fuzzy set G, the GRAY LEVEL ESTIMATOR fuzzy set, to each pair of fuzzy sets EG and TG,

$$G = F(EG, TG) \tag{5.102}$$

in the following way: since any fuzzy set is fully defined by specifying its membership function, we will define the set mapping F in terms of real-valued mappings between the parameters of the membership function $\varphi_G(\widetilde{f}_{i_1,i_2})$ and the parameters of the membership functions $\varphi_{EG}(\widetilde{f}_{i_1,i_2})$ and $\varphi_{TG}(\widetilde{f}_{i_1,i_2})$ as follows:

$$\varphi_G(\widetilde{f}_{i_1,i_2}) = e^{-\xi_G(\widetilde{f}_{i_1,i_2} - \widetilde{f}_{i_1,i_2}^G)^2} \tag{5.103}$$

$$\widetilde{f}_{i_1,i_2}^G \equiv \widetilde{\mu}_E(\kappa_{i_1,i_2})\widetilde{f}_{i_1,i_2}^{(6)} + \widetilde{\mu}_T(\kappa_{i_1,i_2})\widetilde{f}_{i_1,i_2}^{(7)} \tag{5.104}$$

$$\xi_G \equiv \widetilde{\mu}_E(\kappa_{i_1,i_2})\xi_{EG} + \widetilde{\mu}_T(\kappa_{i_1,i_2})\xi_{TG} \tag{5.105}$$

under the condition $\widetilde{f}_{i_1,i_2} \geq \overline{\widetilde{f}}_{i_1,i_2}$. For $\widetilde{f}_{i_1,i_2} < \overline{\widetilde{f}}_{i_1,i_2}$, the terms $\widetilde{f}_{i_1,i_2}^{(6)}$ and $\widetilde{f}_{i_1,i_2}^{(7)}$ are replaced by $\widetilde{f}_{i_1,i_2}^{(4)}$ and $\widetilde{f}_{i_1,i_2}^{(3)}$ respectively.

The mapping operation is illustrated in Figure 5.10. From the figure, it is seen that the mapping performs a kind of interpolation between the two fuzzy sets EG and TG. The coefficients $\widetilde{\mu}_E(\kappa_{i_1,i_2})$ and $\widetilde{\mu}_T(\kappa_{i_1,i_2})$ refer to the previous normalized fuzzy coefficient values of the fuzzy ETC measure model, which is independent of the current gray level estimator fuzzy model. If the current pixel belongs to an edge, the conditions $\widetilde{\mu}_E(\kappa) \approx 1$ and $\widetilde{\mu}_T(\kappa) \approx 0$ are approximately satisfied, the resulting final gray level estimator fuzzy set G is approximately equal to the edge gray level estimator fuzzy set EG, and the final prediction for the gray level value, $\widetilde{f}_{i_1,i_2}^d$, will be based on the corresponding membership

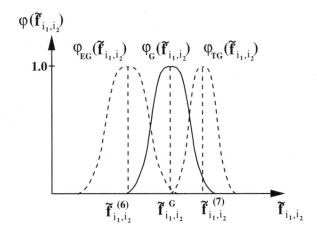

Figure 5.10: The mapping of the fuzzy sets EG and TG to G

function of this fuzzy set. On the other hand, if the current pixel belongs to a textured area, the conditions $\widetilde{\mu}_E(\kappa_{i_1,i_2}) \approx 0$ and $\widetilde{\mu}_T(\kappa_{i_1,i_2}) \approx 1$ hold approximately, and the estimation of $\widetilde{f}^d_{i_1,i_2}$ will be mostly based on the membership function of the texture gray level estimator fuzzy set TG. For all the other cases, the mapping operation results in a membership function for the set G with values of parameters intermediate between those of EG and TG.

5.15.3   Defuzzification of the Fuzzy Set G

The fuzzy inference procedure determines the final GRAY LEVEL ESTIMATOR fuzzy set G with corresponding membership function $\varphi_G(\widetilde{f}_{i_1,i_2})$ from the membership functions $\varphi_{EG}(\widetilde{f}_{i_1,i_2})$ and $\varphi_{TG}(\widetilde{f}_{i_1,i_2})$. In order to provide a desired network output for the fuzzy HMBNN, we have to *defuzzify* the fuzzy set G to obtain a crisp prediction $\widetilde{f}^d_{i_1,i_2}$ and the associated desired network output $\triangle \widetilde{f}^d_{i_1,i_2}$. A common way to defuzzify a fuzzy set is to employ *centroid defuzzification* [37], where we assign the *centroid* of the fuzzy membership function to be the crisp value associated with the function. In the case of the fuzzy set G, we obtain

$$\widetilde{f}^d_{i_1,i_2} \equiv \frac{\int_{-\infty}^{\infty} \widetilde{f} \varphi_G(\widetilde{f}) d\widetilde{f}}{\int_{-\infty}^{\infty} \varphi_G(\widetilde{f}) d\widetilde{f}} = \widetilde{f}^G_{i_1,i_2} \tag{5.106}$$

Due to the adoption of the Gaussian function for $\varphi_G(\widetilde{f}_{i_1,i_2})$, this particular defuzzification strategy does not take into account the information provided by the width of the membership function. Specifically, rather than simply using $\widetilde{f}^G_{i_1,i_2}$, which may not correspond to any of the pixel gray level values in the prediction neighborhood set, as an estimator for the current pixel value, we would also like to incorporate those pixels in the neighborhood set with values "close" to $\widetilde{f}^G_{i_1,i_2}$, in a sense depending on the exact value of the width parameter ξ_G, in the determination of the final estimate for the current gray level value. The inclusion of these pixels from the neighborhood set will reduce incompatibility of the value of the estimator with those pixel values in its neighborhood. In accordance with these, we propose the following *discrete defuzzification procedure* for the fuzzy set G:

$$\widetilde{f}^d_{i_1,i_2} \equiv \widetilde{\varphi}_G(\widetilde{f}^G_{i_1,i_2})\widetilde{f}^G_{i_1,i_2} + \sum_{n_p=1}^{9} \widetilde{\varphi}_G(\widetilde{f}^{(n_p)}_{i_1,i_2})\widetilde{f}^{(n_p)}_{i_1,i_2} \tag{5.107}$$

where $\widetilde{\varphi}_G(\widetilde{f}_{i_1,i_2})$ is the scaled membership function defined as follows:

$$\widetilde{\varphi}_G(\widetilde{f}_{i_1,i_2}) \equiv \frac{1}{C}\varphi_G(\widetilde{f}_{i_1,i_2}) \tag{5.108}$$

$$C \equiv \varphi_G(\widetilde{f}^G_{i_1,i_2}) + \sum_{n_p=1}^{9} \varphi_G(\widetilde{f}^{(n_p)}_{i_1,i_2}) \tag{5.109}$$

From this expression, it is seen that the final crisp estimate $\widetilde{f}^d_{i_1,i_2}$ does not necessarily equal to the centroid $\widetilde{f}^G_{i_1,i_2}$ as in continuous defuzzification, but it also depends on the pixel values in the prediction neighborhood set $\mathcal{N}_p$. In particular, if there exists a pixel with value $\widetilde{f}^{(n_p)}_{i_1,i_2}$ in $\mathcal{N}_p$ which is not exactly equal to $\widetilde{f}^G_{i_1,i_2}$, but is nevertheless "close" to it in the form of a large membership value $\widetilde{\varphi}_G(\widetilde{f}^{(n_p)}_{i_1,i_2})$ in the fuzzy set G, then the crisp estimate $\widetilde{f}^d_{i_1,i_2}$ will be substantially influenced by the presence of this pixel in $\mathcal{N}_p$ such that the final value will be a compromise between $\widetilde{f}^G_{i_1,i_2}$ and $\widetilde{f}^{(n_p)}_{i_1,i_2}$. This is to ensure that, apart from using $\widetilde{f}^G_{i_1,i_2}$, we incorporate the information of all those pixels in the prediction neighborhood set which are close in values to $\widetilde{f}^G_{i_1,i_2}$ to obtain a more reliable estimate of the current pixel value.

5.15.4   Regularization Parameter Update

Given the final crisp estimate $\widetilde{f}^d_{i_1,i_2}$ for the current pixel, we can then define the desired network output $\triangle\widetilde{f}^d_{i_1,i_2}$ for the fuzzy HMBNN:

$$\triangle\widetilde{f}^d_{i_1,i_2} = \widetilde{f}^d_{i_1,i_2} - \widetilde{f}_{i_1,i_2} \tag{5.110}$$

Adopting the same cost function $\mathcal{E}_t$, where

$$\mathcal{E}_t = \frac{1}{2}(\triangle\widetilde{f}^d_{i_1,i_2} - \triangle\widetilde{f}_{i_1,i_2})^2 \tag{5.111}$$

we can update the value of the regularization parameter by applying stochastic gradient descent on this cost function:

$$\lambda_r(t+1) = \lambda_r(t) - \eta\frac{\partial\mathcal{E}_t}{\partial\lambda_r} \tag{5.112}$$

However, unlike the previous case, the fuzzy regularization network has two neurons, namely, the texture neuron and the edge neuron, associated with a single sub-network for the combined edge/textured region, whereas there is only one neuron associated with a sub-network for our previous network. We must therefore derive two update equations for the two regularization parameters $\lambda_{r_f}^{tex}(t)$ and $\lambda_{r_f}^{edge}(t)$ associated with the respective neurons. In addition, each update equation is to be designed in such a way that the resulting value of the regularization parameter would be appropriate for the particular types of regions assigned to each neuron. For example, we may expect that the parameter $\lambda_{r_f}^{tex}(t)$ of the texture neuron would be smaller than the parameter $\lambda_{r_f}^{edge}(t)$ of the edge neuron due to the better noise masking capability of textures.

In view of these considerations, we can formulate the two update equations by regarding the current sub-network as a *two-layer network*. More precisely, we consider the form of equation (5.88), which is repeated here for convenience

$$\triangle \widetilde{f}_{i_1,i_2} = \widetilde{\mu}_E(\kappa_{i_1,i_2})\triangle \widetilde{f}_{i_1,i_2}^{edge} + \widetilde{\mu}_T(\kappa_{i_1,i_2})\triangle \widetilde{f}_{i_1,i_2}^{tex}$$

where $\widetilde{\mu}_E(\kappa_{i_1,i_2}),\widetilde{\mu}_T(\kappa_{i_1,i_2})$ can be regarded as the *output weights* of the subnetwork, and $\triangle \widetilde{f}_{i_1,i_2}^{edge}, \triangle \widetilde{f}_{i_1,i_2}^{tex}$ can be considered as the outputs of the hidden edge and texture neurons. As a result, we can derive the update equations for both $\lambda_{r_f}^{edge}$ and $\lambda_{r_f}^{tex}$ using the *generalized delta rule* for multilayer neural networks.

For the parameter $\lambda_{r_f}^{edge}$ associated with the edge neuron, the update equation is derived as follows:

$$\lambda_{r_f}^{edge}(t+1) = \lambda_{r_f}^{edge}(t) - \eta \frac{\partial \mathcal{E}_t}{\partial \lambda_{r_f}^{edge}}$$

$$= \lambda_{r_f}^{edge}(t) - \eta \frac{\partial \mathcal{E}_t}{\partial \triangle \widetilde{f}_{i_1,i_2}^{edge}} \frac{\partial \triangle \widetilde{f}_{i_1,i_2}^{edge}}{\partial \lambda_{r_f}^{edge}}$$

$$= \lambda_{r_f}^{edge}(t) + \eta(\triangle \widetilde{f}_{i_1,i_2}^{d} - \triangle \widetilde{f}_{i_1,i_2}) \frac{\partial \triangle \widetilde{f}_{i_1,i_2}}{\partial \triangle \widetilde{f}_{i_1,i_2}^{edge}} \frac{\partial \triangle \widetilde{f}_{i_1,i_2}^{edge}}{\partial \lambda_{r_f}^{edge}}$$

$$= \lambda_{r_f}^{edge}(t) + \eta(\triangle \widetilde{f}_{i_1,i_2}^{d} - \triangle \widetilde{f}_{i_1,i_2})\widetilde{\mu}_E(\kappa_{i_1,i_2}) \frac{\partial \triangle \widetilde{f}_{i_1,i_2}^{edge}}{\partial \lambda_{r_f}^{edge}} \qquad (5.113)$$

Similarly, for $\lambda_{r_f}^{tex}$, we have the following update equation:

$$\lambda_{r_f}^{tex}(t+1) = \lambda_{r_f}^{tex}(t) + \eta(\triangle \widetilde{f}_{i_1,i_2}^{d} - \triangle \widetilde{f}_{i_1,i_2})\widetilde{\mu}_T(\kappa_{i_1,i_2}) \frac{\partial \triangle \widetilde{f}_{i_1,i_2}^{tex}}{\partial \lambda_{r_f}^{tex}} \qquad (5.114)$$

From the equations, it is seen that if the current pixel belongs to a textured region, the conditions $\widetilde{\mu}_T(\kappa_{i_1,i_2}) \approx 1$ and $\widetilde{\mu}_E(\kappa_{i_1,i_2}) \approx 0$ are approximately satisfied, and we are essentially updating only the parameter of the texture neuron. On the other hand, if the pixel belongs to an edge, we have the conditions $\widetilde{\mu}_E(\kappa_{i_1,i_2}) \approx 1$ and $\widetilde{\mu}_T(\kappa_{i_1,i_2}) \approx 0$, and we are updating the edge neuron almost exclusively.

5.15.5   Update of the Estimator Fuzzy Set Width Parameters

Previously, it is seen that the width parameters of the gray level estimator fuzzy sets, ξ_{EG} and ξ_{TE}, determine to what degree the various pixels in the prediction neighborhood set participate in estimating the final predicted gray level value. In other words, the parameters establish the notion of "closeness" in gray level values for the various edge/textured regions, which may differ from one such region to another.

In general, we have no *a priori* knowledge of the values of the width parameters required for each combined edge/textured region. In view of this, we propose a simple learning strategy which allows the parameters of each such region to be determined adaptively. Recall that, for the fuzzy sets EG and TG, we assign a membership value of 1 to the order-statistics $\widetilde{f}_{i_1,i_2}^{(6)}$ and $\widetilde{f}_{i_1,i_2}^{(7)}$, respectively (assuming $\widetilde{f}_{i_1,i_2} > \widetilde{f}_{i_1,i_2}$), indicating that these two values are highly relevant in the estimation of the current gray level value. Since the shape of the Gaussian membership function can be fixed by two sample points, we can determine the width parameter by choosing a gray level value in the set of order statistics which is the *least* relevant in estimating the final gray level value, and assign a small membership value $\epsilon \ll 1$ to this gray level value, thus completely specifying the entire membership function.

A suitable candidate for this particular gray level value is the median, or alternatively, the fifth order statistic $\widetilde{f}_{i_1,i_2}^{(5)}$. If we adopt this value as our estimate for the final gray level value, we are effectively aiming at a median filtered version of the image. Although median filter is known to preserve edges while eliminating impulse noises for the case of images without blur [104], we can expect that, for blurred or partially restored images, the median will be close to the mean gray level value in a local neighborhood, and thus would constitute an unsuitable estimate for the final gray level value in the vicinity of edges and textures. We can thus assign a small membership value to the median gray level value to completely determine the final form of the Gaussian membership function.

More precisely, we define the following error function $\mathcal{C}_\xi$ for adapting the two width parameters ξ_{EG} and ξ_{TG} as follows

$$\mathcal{C}_\xi = \frac{1}{2}(\epsilon - \varphi_G(\widetilde{f}_{i_1,i_2}^{(5)}))^2 \qquad (5.115)$$

which expresses the requirement that the value of $\varphi_G(\widetilde{f}_{i_1,i_2})$ should be small when $\widetilde{f}_{i_1,i_2}$ is close to the median value. The gradient descent update equations for the two parameters ξ_{EG} and ξ_{TG} with respect to $\mathcal{C}_\xi$ can then be derived by applying the chain rule according to equations (5.103) and (5.105). For the parameter ξ_{EG},

$$\xi_{EG}(t+1) = \xi_{EG}(t) - \eta_\xi \, \frac{\partial \mathcal{C}_\xi}{\partial \xi_{EG}}$$

$$= \xi_{EG}(t) - \eta_\xi \, \frac{\partial \mathcal{C}_\xi}{\partial \varphi_G} \frac{\partial \varphi_G}{\partial \xi_G} \frac{\partial \xi_G}{\partial \xi_{EG}} \qquad (5.116)$$

Similarly, the update equation for ξ_{TG} is

$$\xi_{TG}(t+1) = \xi_{TG}(t) - \eta_\xi \tilde{\mu}_T(\kappa_{i_1,i_2}) \varphi_G(\tilde{f}_{i_1,i_2}^{(5)})(\epsilon - \varphi_G(\tilde{f}_{i_1,i_2}^{(5)}))(\tilde{f}_{i_1,i_2}^{(5)} - \tilde{f}_{i_1,i_2}^G) \quad (5.117)$$

From the form of the two update equations, it is seen that, similar to the case for the update of the regularization parameters, when the current pixel belongs to a textured region, where $\tilde{\mu}_T(\kappa_{i_1,i_2}) \approx 1$ and $\tilde{\mu}_E(\kappa_{i_1,i_2}) \approx 0$, substantial update is performed on ξ_{TG} while there is essentially no update on ξ_{EG}, which is reasonable in view of the necessity to update the shape of the TG membership function using information from the textured pixels only. On the other hand, for edge pixels where $\tilde{\mu}_E(\kappa_{i_1,i_2}) \approx 1$ and $\tilde{\mu}_T(\kappa_{i_1,i_2}) \approx 0$, only the parameter ξ_{EG} will be substantially modified.

5.16 Experimental Results

The fuzzy modular MBNN was applied to three images, including the Lena image, and two images showing a flower and an eagle which are shown in Figure 5.11. For comparison purpose, the non-adaptive restoration results using the Wiener filter and Hopfield restoration network [135] proposed in [46, 105] are included. In addition, we have also included results using conventional adaptive regularization approaches described in [27, 28], where the original isotropic Euclidean norm in the restoration cost function is replaced by a weighted norm as follows

$$E = \frac{1}{2}\|\mathbf{y} - \mathbf{H}\hat{\mathbf{f}}\|_{\mathbf{A}(\hat{\mathbf{f}})}^2 + \frac{1}{2}\lambda(\hat{\mathbf{f}})\|\mathbf{D}\hat{\mathbf{f}}\|_{\mathbf{B}(\hat{\mathbf{f}})}^2 \quad (5.118)$$

The diagonal weighting matrix $\mathbf{B}(\hat{\mathbf{f}})$ allows the selective emphasis of either the data or regularization term in specific image pixel subsets to achieve spatially adaptive image restoration. Specifically, $\mathbf{B}(\hat{\mathbf{x}})$ is defined such that the associated matrix entries $b_{ii}, i = 1, \ldots, N_I$ are decreasing functions of the local variances, which in turn allows enforcement of the smoothness constraint in smooth regions for noise suppression, and relaxation of this constraint in high variance regions for feature enhancement. In the experiments, the following form of $\mathbf{B}$, proposed in [28], was adopted.

$$b_{ii}(\hat{\sigma}_i) = \frac{1}{1 + \xi \max(0, \hat{\sigma}_i^2 - \sigma_n^2)} \quad (5.119)$$

with $\hat{\sigma}_i^2$ representing the local variance at the ith pixel, σ_n^2 denoting the additive noise variance, and ξ is a tunable parameter. In order to provide a certain degree of adaptivity with respect to varying noise levels for this approach, we specify ξ to be equal to $\frac{1}{\sigma_n^2}$. The role of the weighting matrix $\mathbf{A}(\hat{\mathbf{f}})$ is complementary to that of $\mathbf{B}(\hat{\mathbf{f}})$ and is defined as $\mathbf{I} - \mathbf{B}(\hat{\mathbf{f}})$.

The global regularization parameter in equation (5.118) is defined as the following function of the partially restored image $\hat{\mathbf{f}}$ as proposed in [27, 113]

$$\lambda(\hat{\mathbf{f}}) = \frac{\|\mathbf{y} - \mathbf{H}\hat{\mathbf{f}}\|_{\mathbf{A}(\hat{\mathbf{f}})}^2}{2\|\mathbf{y}\|^2 - \|\mathbf{D}\hat{\mathbf{f}}\|_{\mathbf{B}(\hat{\mathbf{f}})}^2} \quad (5.120)$$

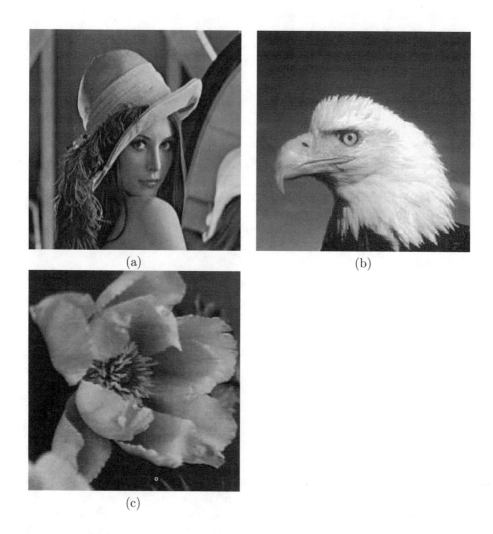

(a)   (b)

(c)

Figure 5.11: Original images. (a) Lena (b) Eagle (c) Flower

In Figure 5.12, a 5×5 point spread function implementing a uniform blur is applied to the image Lena. In addition, Gaussian noise at the level of 30 dB BSNR (Blurred Signal-to-Noise ratio) [23] is added to the image. The degraded image is shown in Figure 5.12a. The restoration result using the current fuzzy MBNN restoration algorithm is shown in Figure 5.12e. Figure 5.12b shows the result using the Wiener filter, where we can notice the blurring and severe ringing at the boundaries. Figure 5.12c shows the result using the non-adaptive Hopfield network restoration approach proposed by Zhou *et al.* [46], where the global regularization parameter λ is determined as $\lambda = 1/\text{BSNR}$ according to [22]. We can notice the noisy appearance of the image due to the non-adaptive regularization approach. Figure 5.12d shows the result using the parameter assignment function (5.119) with $\xi = 1/\sigma_n^2$. In addition, the global parameter $\lambda(\hat{\mathbf{x}})$ is updated iteratively according to (5.120). Although Figure 5.12d is an improvement on Figure 5.12b and c, the ringing in the vicinity of the edges is still noticeable.

The above results can be compared with the restoration result using the current approach in Figure 5.12e. We can notice that noise suppression around the edges is achieved without compromising the details of the textured area.

The importance of adopting different regularization parameters for edges and textures is illustrated in Figure 5.13. In Figure 5.13a, we magnify the lower left portion of the Lena image to show the effect of adopting the texture-oriented regularization (5.98) to edges. The noisy appearance of the restored edge is readily noticed when this is compared with the same region restored using edge-oriented regularization in Figure 5.13b. On the other hand, if we apply an edge-oriented regularization approach to textures as shown in Figure 5.13c, we can notice the blurred appearance which can be compared with the same region restored using the texture-oriented regularization approach in Figure 5.13d.

Figures 5.14a to e show the restoration results under 5×5 uniform blur with 20 dB BSNR, which represents a more severe degradation for the image. Comparing with the degraded image is shown in Figure 5.14a, we can notice that the Wiener filter restoration result in Figure 5.14b does not result in an appreciable enhancement of image details. This is due to the increased emphasis of the Wiener filter on noise smoothing rather than feature enhancement at higher noise levels. Similarly, we can notice the noisy appearances of the non-adaptive Hopfield network result in Figure 5.14c and the spatially adaptive regularization result in Figure 5.14d. We can adopt alternative values for ξ in equation (5.119) for noise reduction, but there are in general no effective criteria for choosing the optimal parameter value. As a result, to illustrate the best performance under equation (5.119), the parameter ξ is adjusted in such a way that the root mean square (RMSE) between the restored and original image is minimized, which is only possible with the availability of the original image and does not genuinely represent the usual case where only the blurred image is available. In other words, we can expect that the RMSE in practical restoration attempts to be higher. The restored image using this parameter selection approach is shown in Figure 5.14e, with the associated RMSE value shown in Table 5.2. We can see that even with the minimization of RMSE through a suitable choice of ξ does not necessarily result in a restored image with good visual quality, as seen in the

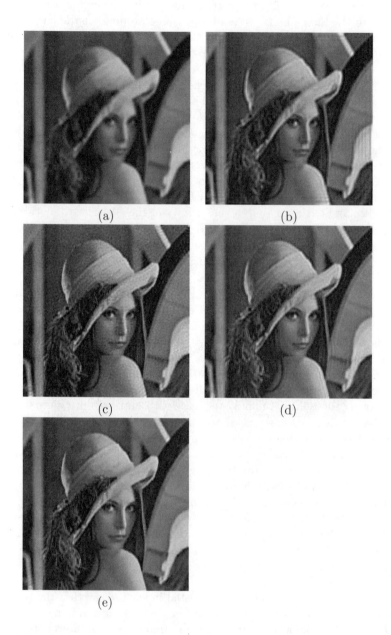

Figure 5.12: Restoration of the Lena image (5×5 uniform blur, 30 dB BSNR).
(a) Degraded image. (b)-(d) Restored images using (b) Wiener filter, (c) Hopfield
NN ($\lambda = 1/BSNR$), (d) Adaptive restoration approach using equation (5.119)
($\xi = 1/\sigma_n^2$), (e) Fuzzy MBNN

Figure 5.13: (a) Edges under texture-oriented regularization. (b) Edges under edge-oriented regularization. (c) Textures under edge-oriented regularization. (d) Textures under texture-oriented regularization

Figure 5.14: Restoration of Lena image (5×5 uniform blur, 20 dB BSNR). (a) Degraded image. (b)-(f) Restored images using (b) Wiener filter, (c) Hopfield NN ($\lambda = 1/BSNR$), (d) Adaptive restoration using equation (5.119)($\xi = 1/\sigma_n^2$), (e) Adaptive restoration using equation (5.119)(optimized ξ), (f) Fuzzy MBNN

Figure 5.15: Further restoration results using fuzzy MBNN. (a) Degraded eagle image (30 dB BSNR). (b) Degraded flower image (20 dB BSNR). (c) Restored eagle image. (d) Restored flower image

blurred appearance of the restored image. This can be compared with the result using the current approach in Figure 5.14f, where it is seen that the visual quality is satisfactory even at this more serious level of degradation. This is achieved by the possibility of distinguishing between various feature types through the ETC measure and the corresponding assignment of different λ_r values to these feature classes.

We have also applied the current algorithm to the eagle and flower images. We apply the uniform PSF with 30 dB BSNR additive Gaussian noise to the eagle image in Figure 5.15a, and the corresponding restored image is shown in Figure 5.15c. Similarly, the flower image is degraded with 5×5 uniform blurring and 20 dB Gaussian additive noise in Figure 5.15b, and the restored image is shown in Figure 5.15d. In general, conclusions similar to those for the previous

Table 5.2: RMSE values of the restoration results using different restoration algorithms

Image,BSNR	Blur.	Wien.	N.adap.	(119)	(119)(opt.ξ)	HMBNN
Lena, 30 dB	13.03	11.23	8.65	8.46	8.45	7.09
Lena, 20 dB	13.79	11.61	10.56	12.62	11.43	9.69
Eagle, 30 dB	9.88	9.61	10.28	7.81	7.62	6.89
Eagle, 20 dB	11.98	9.75	11.57	13.03	11.07	8.99
Flower, 30 dB	8.73	7.96	7.70	6.15	5.90	4.98
Flower, 20 dB	10.06	8.12	8.47	11.00	8.65	6.92

(a)   (b)

Figure 5.16: Distribution of local λ values. (a) and (b) Lena (a) 30 dB, (b) 20 dB

cases can be drawn from these results with regard to the restoration qualities in both the edge and textured regions. Comparison of the restoration results for these two images in terms of the root mean square (RMSE) measure is also shown in Table 5.2.

The spatial distribution of regularization parameter values for the Lena image are shown in Figure 5.16. In these images, light gray level values correspond to large regularization parameter values λ_r, and dark areas correspond to small values of λ_r. The λ distribution corresponding to 5 $\times$ 5 times uniform blur at 30 dB BSNR are shown in Figure 5.16a, while those corresponding to the same PSF at 20 dB BSNR are shown in Figure 5.16b. It is seen that, for smooth regions, large λ_r values are generated for greater noise suppression, while small values are automatically adopted for high variance regions to achieve feature enhancement. In addition, it is seen that the λ maps for the 20 dB case exhibit a lighter gray values than those for the 30 dB case, which is reasonable considering the greater

(a)   (b)

Figure 5.17: (a) and (b) Lena (a) Edge pixels, (b) Texture pixels

need of noise suppression in this case. Thus it is seen that, instead of attempting
to assign regularization parameters directly, the optimal values of which may
depend on the particular noise levels, it is more desirable to adopt a learning
approach where the specification of a single gray level target value for the fuzzy
NN allows the emergence of different local regularization parameters in response
to different noise levels.

In Figure 5.17, we show the thresholded distribution map of ETC values for
the Lena image. Specifically, in Figure 5.17b, those pixels with their ETC coef-
ficient values $\widetilde{\mu}_T(\kappa_{r,i}) > \widetilde{\mu}_E(\kappa_{r,i})$ are labeled. We should expect that, according
to our previous discussion, these pixels will approximately correspond to the tex-
tures and be regularized essentially by the smaller λ_r^{tex}. This can be confirmed
from the figures where it is seen that the labeled pixels are mostly clustered
around the feathers on the hat. In Figure 5.17a, we have labeled those image
pixels with $\widetilde{\mu}_E(\kappa_{r,i}) \geq \widetilde{\mu}_T(\kappa_{r,i})$. They will essentially be regularized with the
larger parameter λ_r^{edge} and thus we expect they should approximately correspond
to the edges, which is again confirmed from the figures.

In addition to the visual comparisons, we have included objective comparisons
in terms of the root mean square error (RMSE) measure, which is defined as
follows:

$$RMSE = \left(\frac{1}{N_I} \| \mathbf{f} - \hat{\mathbf{f}} \|^2 \right)^{\frac{1}{2}} \tag{5.121}$$

where $\mathbf{f}$ and $\hat{\mathbf{f}}$ represent the original image and restored image, respectively, with
the pixels arranged in lexicographic order, and N_I is the number of pixels in the
image. The RMSE values of the restored images using the various algorithms
are listed in Table 5.2. The improvement resulting from the adoption of the
current approach is indicated by the corresponding small RMSE values at various
noise levels. In particular, we can compare the RMSE values of the current

approach with those using the adaptive strategy (5.119) with optimized ξ with respect to the *original image* in column 5. Although in some cases (especially under low noise levels), these values are only slightly higher than those using the current approach, we can expect that in practical cases where the original image is not available, the corresponding RMSE values will be higher than those shown. In addition, at higher noise levels, the RMSE values using the adaptive approach (5.119) is significantly higher than those using the current approach.

5.17   Summary

An alternative formulation of problem of adaptive regularization in image restoration was proposed in the form of a *model-based neural network with hierarchical architecture* (HMBNN) operating on non-overlapping regions on the image. Based on the principle of adopting small parameter values for the textured regions for detail emphasis while using large values for ringing and noise suppression in the smooth regions, we have developed an alternative viewpoint of adaptive regularization which centers on the concept of *regularization parameters as model-based neuronal weights*. From this we have derived a stochastic gradient descent algorithm for optimizing the parameter value in each region. In addition, incremental re-definition of the various regions using the nearest neighbor principle is incorporated to reverse the effects of initial inaccurate segmentation.

We have generalized the previous HMBNN framework for adaptive regularization to incorporate fuzzy information for edge/texture discrimination. Adopting the new ETC measure, we propose an edge/texture fuzzy model which expresses the degree to which a local image neighborhood resembles either edge or texture in terms of two fuzzy membership values. Correspondingly, we modify our previous network architecture to incorporate two neurons, namely, the edge neuron and the texture neuron, within each sub-network. Each of the two neurons estimate an independent regularization parameter from the local image neighborhood and evaluate the corresponding required gray level update value as a function of its own parameter. The two gray level update values are then combined using fuzzy inference to produce the final required update in a way which takes into account the degree of edge/texture resemblance of the local neighborhood. In general, the algorithm is designed such that less regularization is applied to the textured areas due to their better noise masking capability, and more regularization is applied to the edges where the noises are comparatively more visible.

The generalized algorithm is applied to a number of images under various conditions of degradations. The better visual quality of the restored images under the current algorithm can be appreciated by comparing the result with those produced using a number of conventional restoration algorithms. The current fuzzy HMBNN paradigm thus refines the previous notion of simply applying less regularization for the combined edge/textured regions to allow the possibility of using different levels of regularization to accommodate the different noise masking capabilities of the various regional components.

Chapter 6

Adaptive Regularization Using Evolutionary Computation

6.1 Introduction

In this chapter, we propose an alternative solution to the problem of adaptive regularization by adopting a new *global* cost measure. It was previously seen that the newly formulated ETC measure is capable of distinguishing between the textures and edges in an image. It is further observed in this chapter that the distribution function of this measure value in a typical image assumes a characteristic shape, which is comparatively invariant across a large class of images, and can thus be considered a *signature* for the images. This is in contrast with the distribution function of other quantities such as the gray level values which varies widely from image to image, as can be confirmed by observing the gray level histograms of different images.

It is also observed that the corresponding ETC distribution function (or equivalently the ETC probability density function (ETC-pdf)) of degraded images is usually very different from that of the non-degraded images. In other words, for optimal restoration results, we can assign the regularization parameters in such a way that the ETC distribution function of the restored image once again assumes the shape which is typical for non-degraded images, or more formally, we have to minimize the distance between the ETC-pdf of the restored image and the characteristic ETC-pdf of non-degraded images.

In practice, we can only approximate the ETC-pdf by the histogram of the measure values in the image, which cannot be expressed in closed form with respect to the regularization parameters, and thus the conventional gradient-based optimization algorithms are not applicable. We have therefore adopted an artificial evolutionary optimization approach where evolutionary programming (EP), belonging to the class of algorithms known as evolutionary computational algorithms, is used to search for the optimal set of regularization parameters with respect to the ETC-pdf criterion. One of the advantages of this class of algorithms is their independence from the availability of gradient information, which is therefore uniquely suited to our current optimization problem. In addition, these algorithms employ multiple search points in a population instead of a sequence of single search points as in conventional optimization algorithms, thus allowing many regions of the parameter space to be explored simultaneously. The characteristics of this class of algorithms are described in the following section.

6.2   Introduction to Evolutionary Computation

Evolutionary programming [40, 136] belongs to the class of optimization algorithms known as evolutionary computational algorithms [39, 40, 137, 138, 139] which mimic the process of natural evolution to search for an optimizer of a cost function. There are three mainstreams of research activities in this field, which include research in genetic algorithm (GA) introduced by Holland, Goldberg and Mitchell [61, 140, 141], evolutionary programming (EP) by Fogel [40, 136], and evolutionary strategy (ES) by Rechenberg and Schwefel [142, 143]. The defining characteristic of this class of algorithms includes its maintenance of a diversity of potential optimizers in a *population* and allow highly effective optimizers to emerge through the processes of mutation, recombination, competition and selection. The implicit parallelism resulting from the use of multiple search points instead of a single search point in conventional optimization algorithms allows many regions of the search space to be explored simultaneously. Together with the stochastic nature of the algorithms which allow search points to spontaneously escape from non-global optima, and the independence of the optimization process from gradient information, the instances of local minima are usually reduced, as unlike the case with the gradient-based algorithms. In addition, this independence from gradient information allows the incorporation of highly irregular functions as fitness criteria for the evolutionary process, unlike the case with gradient-based algorithms where only differentiable cost functions are allowed. It is not even necessary for the cost function to have a closed form, as in those cases where the function values are obtained only through simulation. We will give a brief introduction of the above members of this class of algorithms in the following sections.

6.2.1   Genetic Algorithm

Genetic algorithm (GA) [61, 140, 141] is the most widely used among the three evolutionary computational algorithms. The distinguishing feature of GA includes its representation of the potential optimizers in the population as *binary strings*. Assuming the original optimizers are real-valued vectors $\mathbf{z} \in \mathbf{R}^N$, an encoding operation $\mathcal{G}$ is applied to each of these vectors to form the binary strings $\mathbf{g} = \mathcal{G}(\mathbf{z}) \in \mathbf{B}^H$, where $\mathbf{B} = \{0, 1\}$. In other words, the various evolutionary operations are carried out in the space of *genotypes*.

New individuals in the population are created by the operations of *crossover* and *mutation*. In GA, crossover is the predominant operation, while mutation only serves as an infrequent background operation. In crossover, two binary strings $\mathbf{g}_{p_1}, \mathbf{g}_{p_2}$ are randomly selected from the population. A random position $h \in \{1, \ldots, H\}$ is selected along the length of the binary string. After this, the sub-string to the left of h in p_1 is joined to the right sub-string of p_2, and similarly for the left sub-string of p_2 and the right sub-string of p_1, thus mimicking the biological crossover operation on the chromosomes. On the other hand, the mutation operator toggles the status of each bit for a certain binary string with a probability π_m, which is a very small value in the case of GA. The main purpose of mutation is to introduce new variants of genotypes into the population.

After the processes of crossover and mutation, the fitness of each binary string, $f(\mathbf{g}_p), p = 1, \ldots, \mu$, is evaluated, where μ is the number of optimizers in the population, and f represents the *fitness function*. The function f usually reflects the requirement of the optimization problem at hand. In the case of a maximization task, we can usually equate the objective function of the problem with the fitness function. In the case of a problem which requires minimization, we can simply set the fitness function to be the negative of the current cost function.

After the evaluation of the individual fitness values, the optimizers in the population undergo a proportional selection process: each optimizer is to be included in the next generation with probability $\pi(\mathbf{g}_{p'})$, which reflects the relative fitness of individual $\mathbf{g}_{p'}$

$$\pi(\mathbf{g}_{p'}) = \frac{f(\mathbf{g}_{p'})}{\sum_{p=1}^{\mu} f(\mathbf{g}_p)} \tag{6.1}$$

In other words, an individual is more likely to survive into the next generation if it possesses a high fitness value. As the algorithm proceeds, the population will eventually consist of those optimizers with appreciable fitness values.

6.2.2 Evolutionary Strategy

As opposed to genetic algorithm, evolutionary strategy (ES) [142, 143] represents an alternative evolutionary approach where the various adaptation operations are carried out in the space of *phenotypes*: instead of first encoding the individual optimizers $\mathbf{z} \in \mathbf{R}^N$ into binary strings, the *recombination* and *mutation* operations are carried out directly on the real-valued vectors as described below.

For the recombination operation, two optimizers $\mathbf{z}_{p_1}, \mathbf{z}_{p_2}$ are randomly selected from the population, and a new optimizer $\mathbf{z}$ is generated from these two according to the recombination operator C.

$$\mathbf{z} = C(\mathbf{z}_{p_1}, \mathbf{z}_{p_2}) \tag{6.2}$$

The simplest form for C is the linear combination operation

$$\mathbf{z} = \alpha \mathbf{z}_{p_1} + (1 - \alpha) \mathbf{z}_{p_2} \tag{6.3}$$

where $\alpha < 1$ is the combination coefficient, although other combination operations, as described in [39], are also possible.

The mutation operation in ES randomly perturbs each component of the optimizer vector z_j to form a new optimizer $\mathbf{z}'$ with components z'_j as follows

$$z'_j = z_j + N(0, \sigma_j^m) \qquad (6.4)$$

where $N(0, \sigma)$ is a Gaussian random variable with mean 0 and standard deviation σ. In the terminology of ES, the parameter σ_j^m associated with the component z_j is usually referred to as a mutation *strategy parameter*. The values of the mutation strategy parameters determine whether the current optimization process more resembles a global search, as when the σ_j^m's assume large values, which is desirable at the initial stages when many regions of the parameter space are simultaneously explored, or a local search which is more appropriate toward the final stages when the mutation strategy parameters assume small values to restrict the search within promising localities. The mutation strategy parameters themselves are usually adapted according to the following log-normal equation:

$$\sigma'^m_j = \sigma_j^m \exp(\tau' N(0,1) + \tau N_j(0,1)) \qquad (6.5)$$

where τ', τ are pre-determined constants, and $N(0,1), N_j(0,1)$ are Gaussian random variables with mean 0 and standard deviation 1. In the case of $N_j(0,1)$, the random variable is re-sampled for each new component j. The log-normal adaptation equation is adopted to preserve the positivity of the mutation strategy parameters σ_j^m.

The recombination and mutation operations are performed γ times to form γ new individuals from the original μ individuals in the population. In the case of $(\mu + \gamma)$ selection strategy [39], the fitness values $f(\mathbf{z}_p)$ of each individual in the $(\mu + \gamma)$ parent/descendant combination are evaluated, and those μ optimizers in this combination with the greatest fitness values are incorporated into the population in the next generation. In the (μ, γ) selection strategy [39], only the fitness values of the newly generated γ descendants are evaluated and the fittest of those incorporated into the next generation. The selection process is deterministic and depends solely on the fitness values of the individuals.

6.2.3   Evolutionary Programming

Evolutionary Programming (EP) [40, 136] shares many common features with evolutionary strategy in that the primary adaptation operations are also carried out in the space of phenotypes. In addition, there are a number of important similarities.

- The individual components z_j of each optimizer $\mathbf{z} \in \mathbf{R}^N$ are also perturbed according to equation (6.4), with the mutation strategy parameters σ_j^m similarly defined for the current component. In some variants of EP, the individual components are perturbed by an amount proportional to the square root of the objective function value [40].

- The individual mutation strategy parameters σ_j^m themselves are also subject to adaptations. In some EP variants, the log-normal relationship (6.5) is also adopted for this purpose.

Despite these similarities, there are a number of important differences between EP and ES which clearly distinguishes the former from the latter:

- In EP, mutation is the only adaptation operation applied to the individuals in the population. *No* recombination operations are carried out.

- Instead of using a deterministic selection strategy as in ES, where γ descendant optimizers are created from μ parent optimizers, and the members of the new population selected from the resulting combination according to a ranking of their respective fitness values, EP uses a stochastic selection strategy as follows: for each optimizer in the $(\mu + \gamma)$ parent/descendant combination, we randomly select Q other optimizers in the same combination and compare their fitness values with the current optimizer. The current optimizer is included in the population in the next generation if its fitness value is greater than those of the Q "opponents" selected. In addition, the number of descendants γ is usually set equal to the number of parents μ. In other words, this *Q-tournament selection strategy* can be regarded as a probabilistic version of the $(\mu + \mu)$ selection strategy in ES.

The tournament selection strategy has the advantage that, compared with the deterministic selection strategy, even optimizers which are ranked in the lower half of the $(\mu + \mu)$ parent and descendant combination have a positive probability of being included in the population in the next generation. This is especially useful for non-stationary fitness functions when certain optimizers in the population which at first seem non-promising turn out to be highly relevant due to the constant evolving nature of the fitness landscape. These optimizers may probably be already excluded in the initial stages of optimization if a deterministic strategy is adopted, but in the case of a tournament selection strategy, it is still possible for these optimizers to be included if there are indications that they are slightly distinguished from the truly non-promising optimizers through the result of the tournament competition.

This is also the reason for our choice of EP for our adaptive regularization problem: in our algorithm, we have generated a population of *regularization strategies*, which are vectors of regularization and segmentation parameters, as our potential optimizers. The optimal regularization strategy is selected from the population at each generation according to criteria to be described in a later section, and this is used to restore the image for a single iteration. This partially restored image is then used as the basis for the evolution of regularization strategies in the next generation. In other words, the fitness landscape in the next generation depends on the particular optimizer selected in the previous generation, thus constituting a non-stationary optimization problem. The stochastic tournament selection strategy is therefore useful in retaining those regularization strategies which initially seem non-promising but are actually highly relevant in later restoration stages.

6.3   The ETC-pdf Image Model

In this chapter, we address the adaptive regularization problem by proposing a novel image model, the adoption of which in turn necessitates the use of powerful optimization algorithms such as those typical in the field of evolutionary computation. This model is observed to be capable of succinctly characterizing common properties of a large class of images, and thus we will regard any restored image which conforms to this image model as suitably regularized. In other words, this model can be regarded as a possible objective characterization of our usual notion of subjective quality. The model, which is specified as the probability distribution of the ETC measure introduced in Chapter 5, is approximated as a histogram, and the regularization parameters in the various image regions are chosen in such a way that the corresponding ETC histogram of the resulting restored image matches the model pdf closely, i.e., minimizing the difference between the two distributions. It is obvious that the resulting error function, which involves differences of discrete distributions, is highly irregular and non-differentiable, and thus necessitates the use of powerful optimization algorithms. In view of this, we have chosen evolutionary programming as the optimization algorithm to search for the minimizer of this cost function.

Denoting the probability density function of the ETC measure κ within a typical image as $p_\kappa(\kappa)$, we have plotted the ETC histograms, which are approximations of the ETC-pdf, for several images in Figure 6.1. It is noticed that the histograms peak around $\kappa = 1$, indicating the predominance of smooth regions. As κ increases, values of the various histograms gradually decrease, with $p_\kappa(\kappa) \approx 0$ for $\kappa \approx K$, which is the size of the averaging window used ($K = 5$ in the current example). This indicates the smaller proportion of edges and textured regions. More importantly, it is seen that, although there are slight deviations between the various ETC histograms, they in general assume the typical form as shown in Figure 6.1. This is in contrast with the traditional gray level histogram which is usually used to characterize the gray level distribution in an image, and which can vary widely from image to image. This is illustrated in Figures 6.2a to d, where the gray level histograms for the same images are shown. As a result, we can consider the ETC histogram as a form of *signature* for a large class of non-degraded images.

On the other hand, it is observed that the corresponding density functions for degraded images are usually very different from the standard density function. Figure 6.3 illustrates this point by comparing the corresponding ETC-pdf of one of the images in Figure 6.1 and its blurred version. In the figure, the solid curve is the original ETC-pdf and the dotted curve is the ETC-pdf of the blurred image. It is seen that the rate of decrease is greater for the blurred image, indicating the higher degree of correlation among its pixels. Therefore, one possible regularization strategy to allow the restored image to more closely resemble the original image is to assign the parameters in such a way as to minimize the discrepancies between the ETC histogram of the restored image and that of the original.

Due to the similar shapes of the ETC-pdf in Figure 6.1, we can model the

Figure 6.1: The ETC-pdf of different images

Figure 6.2: Gray level histograms of different images: (a) Image 1, (b) Image 2, (c) Image 3, (d) Image 4

Figure 6.3: The ETC-pdf for a typical image and its blurred version. Solid curve is the ETC-pdf of the original image. Dotted curve is the ETC-pdf of the blurred image. Size of averaging window is 5×5

ETC-pdf using a combination of piecewise Gaussian and exponential functions. During restoration, we can adaptively assign the regularization parameters in such a way that the corresponding ETC-pdf in the restored image conforms closely to the model density function. In this work, we have adopted the following model $p_\kappa^M(\kappa)$ for the typical ETC-pdf, which can best characterize the density function of a large class of images.

$$
p_\kappa^M(\kappa) = \begin{cases} ce^{-(\kappa-1)^2} & 0 \leq \kappa \leq 1 \\ ca_1^{\varepsilon(\kappa-1)} & 1 < \kappa \leq \kappa_3 \\ ca_1^{\varepsilon(\kappa_3-1)} a_2^{\varepsilon(\kappa-\kappa_3)} & \kappa_3 < \kappa \leq K \end{cases} \tag{6.6}
$$

In this equation, we use a Gaussian function segment to model the density function when $\kappa \in [0,1]$, and use two exponential function segments to model the tail distribution in the interval $[1, K]$, with the corresponding parameters a_1 and a_2 satisfying $a_1 < a_2$. The constant κ_3 in the equation is that value of κ which corresponds to an almost equipartition of the K^2 variables into three components. It has been shown for the case of 5×5 averaging window that $\kappa_3 \approx 1.8$. We could have modeled the density function in the interval $[1, K]$ using a single exponential segment with parameter a, and in turn estimate this parameter from a histogram of κ using typical real-world images. Instead, we have used two separate exponential functions, with a_1 and a_2 chosen such that $a_1 < a_2$. The reason for doing this and choosing the particular transition point κ_3 is that: for $\kappa \in [1, \kappa_3]$, the neighborhood surrounding the current pixel consists of fewer than three components, which usually corresponds to a mixture of noises and smooth regions, and is particularly undesirable given the enhanced visibility of noises against the smooth background. One must therefore limit the probability of occurrence of such values of κ, which explains the adoption of $a_1 < a_2$ in the probability density model allowing a smaller probability of occurrences for $\kappa \in [1, \kappa_3]$. One may argue that this may adversely affect the probability of edge occurrence, which consists of two gross components with $\kappa = \kappa_2 \in [1, \kappa_3]$ as well, but edges usually occupy a small area in a typical image, which translates to a very small occurrence probability, so it is much more probable that those locations with κ in that interval correspond to a mixture of noises and smooth backgrounds, and the main effect of probability reduction in this interval is the elimination of this type of artifact. The variable ε controls the rates of decay for the two exponentials, and the constant c in the equation is a normalization factor such that

$$
\int_0^K p_\kappa^M(\kappa) \, d\kappa = 1 \tag{6.7}
$$

6.4   Adaptive Regularization Using Evolutionary Programming

As mentioned previously, the value of the regularization parameter in image restoration applications is usually determined by trial and error. Therefore, the purpose of establishing the model in Section 6.3 is to provide an objective way by which we can assign the parameters adaptively to result in the best subjective quality. Formally, we replace the constrained least square cost function for image restoration with the following cost function:

$$E = \frac{1}{2}\|\mathbf{y} - \mathbf{H}\hat{\mathbf{f}}\|_{\mathbf{A}}^2 + \frac{1}{2}\|\mathbf{D}\hat{\mathbf{f}}\|_{\mathbf{L}}^2 \tag{6.8}$$

where, instead of a single parameter λ, we employ a diagonal *weighting matrix* $\mathbf{L}$ defined as follows:

$$\mathbf{L} = diag[\lambda(\sigma_1), \dots, \lambda(\sigma_{N_I})] \tag{6.9}$$

where $\sigma_i, i = 1, \dots, N_I$ is the local standard deviation of the ith pixel in the image. This is similar to the cost function associated with the spatially adaptive iterative restoration algorithm which was adopted as a benchmark in Chapter 5. However, an important difference between the current formulation and the previous algorithm is that, while the previous algorithm performs adaptive regularization by separately adjusting a global parameter $\lambda(\hat{\mathbf{f}})$ and a weighting matrix $\mathbf{A}$, the current formulation combines these two into a single weighting matrix $\mathbf{L}$, the entries of which are determined through evolutionary programming. Similarly, the weighting matrix $\mathbf{A}$ is usually specified in such a way as to complement the effect of $\mathbf{L}$, or is sometimes simply replaced by the identity matrix $\mathbf{I}$, as we have chosen for the current algorithm.

Denote the ETC-pdf for the restored image $\hat{\mathbf{f}}$ as $p_\kappa(\kappa)$, our objective is to select the particular forms for $\lambda(\sigma_i)$ in the weighting matrix $\mathbf{L}$ in such a way to minimize the following *weighted probability density error measure* of the ETC measure κ (ETC-pdf error measure).

$$E_{pdf}^\kappa = \int_0^K w(\kappa)(p_\kappa^M(\kappa) - p_\kappa(\kappa))^2 \, d\kappa \tag{6.10}$$

where the weighting coefficients are defined as follows

$$w(\kappa) \equiv \frac{1}{\max(p_\kappa^M(\kappa), p_\kappa(\kappa))^2} \tag{6.11}$$

to compensate for the generally smaller contribution of the tail region to the total probability. In practice, we replace the integral operation by a finite summation over a suitable discretization of $[0, K]$, and the density function $p_\kappa(\kappa)$ is approximated using the histogram of κ in the partially restored image.

$$E_{pdf}^\kappa = \sum_{r=1}^d w(r\triangle)(p_\kappa^M(r\triangle) - \hat{p}_\kappa(r\triangle))^2 \tag{6.12}$$

where $\triangle$ is the width of the discretization interval, $\hat{p}_\kappa(r\triangle)$ is the estimated ETC-pdf of κ in terms of its histogram, and d is the number of discretization intervals. Since the histogram records the relative occurrence frequencies of the various values of κ based on the previous discretization, it involves the counting of discrete quantities. As a result, the overall error function (6.12) is obviously non-differentiable with respect to the regularization parameters, and the evolutionary programming approach provides a viable option to minimize this error function.

Evolutionary programming is a population-based optimization algorithm in which the individual optimizers in the population compete against each other with respect to the minimization of a cost function, or equivalently, the maximization of a fitness function [137]. In the current case, we have already specified the fitness function, which is equation (6.12). In addition, we have to specify the form of the individual optimizers in the population as well. For the adaptive regularization problem, we consider the following *regularization profile* $\lambda(\sigma_i)$ defined on the local standard deviation σ_i around the ith pixel

$$\lambda(\sigma_i) = \frac{\lambda_{\max} - \lambda_{\min}}{1 + e^{\beta(\sigma_i - \alpha)}} + \lambda_{\min} \tag{6.13}$$

Equation (6.13) defines a decreasing sigmoidal function on the local standard deviation range of the image, which is consistent with our previous view that large λ is required at low variance pixels to suppress noise and small λ is required at high variance pixels to enhance the features there. There are four parameters in equation (6.13) which determine the overall λ assignment strategy: $\lambda_{\min}$ and $\lambda_{\max}$ represent the minimum and maximum parameter values used, respectively, α represents the offset of the sigmoidal transition from the origin, thus implicitly defining the standard deviation threshold which separates the small variance from the large variance region, and β controls the steepness of the sigmoidal transition. The various parameters of a typical regularization profile are illustrated in Figure 6.4.

Concatenating these four parameters together with the *mutation strategy parameters* $\sigma_{\lambda_{\min}}^m, \sigma_{\lambda_{\max}}^m, \sigma_\alpha^m, \sigma_\beta^m$ (not to be confused with the local standard deviation σ_i of an image) into an 8-tuple, we define the following *regularization strategy* $\mathcal{S}_p$ as the pth potential optimizer in the population.

$$\mathcal{S}_p \equiv (\lambda_{\min,p}, \lambda_{\max,p}, \alpha_p, \beta_p, \sigma_{\lambda_{\min},p}^m, \sigma_{\lambda_{\max},p}^m, \sigma_{\alpha,p}^m, \sigma_{\beta,p}^m) \tag{6.14}$$

Employing the usual operations of evolutionary computational algorithms, which in the case of evolutionary programming is restricted to the mutation operator [40], we generate a population P consisting of μ instances of $\mathcal{S}_p$ in the first generation, and we apply mutation to each of these μ parents to generate μ descendants in each subsequent generation according to the following equations

$$\lambda'_{\min,p} = \lambda_{\min,p} + N(0, \sigma_{\lambda_{\min,p}}^m) \tag{6.15}$$

$$\lambda'_{\max,p} = \lambda_{\max,p} + N(0, \sigma_{\lambda_{\max,p}}^m) \tag{6.16}$$

$$\alpha'_p = \alpha_p + N(0, \sigma_{\alpha,p}^m) \tag{6.17}$$

$$\beta'_p = \beta_p + N(0, \sigma_{\beta,p}^m) \tag{6.18}$$

where $N(0, \sigma)$ denotes a Gaussian random variable with zero mean and standard deviation σ. The variables $\lambda'_{\min,p}$, $\lambda'_{\max,p}$, α'_p and β'_p are the components of

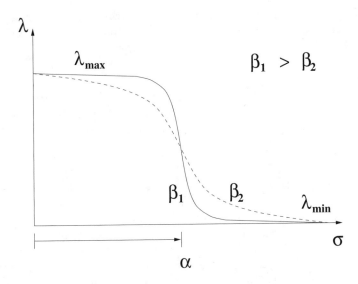

Figure 6.4: Illustrating the various parameters of a typical regularization profile $\mathcal{S}_p$

the descendant strategy $\mathcal{S}'_p$. The mutation strategy parameters are updated according to the following *log-normal adaptation rule*.

$$\sigma'^m_{\lambda_{min},p} = \sigma^m_{\lambda_{min},p} \exp(N(0,\tau_1) + N_j(0,\tau_2)) \qquad (6.19)$$

$$\sigma'^m_{\lambda_{max},p} = \sigma^m_{\lambda_{max},p} \exp(N(0,\tau_1) + N_j(0,\tau_2)) \qquad (6.20)$$

$$\sigma'^m_{\alpha,p} = \sigma^m_{\alpha,p} \exp(N(0,\tau_1) + N_j(0,\tau_2)) \qquad (6.21)$$

$$\sigma'^m_{\beta,p} = \sigma^m_{\beta,p} \exp(N(0,\tau_1) + N_j(0,\tau_2)) \qquad (6.22)$$

where $N(0,\tau_1)$ is held fixed across all mutation parameters, while $N_j(0,\tau_2)$ is generated anew for each parameter. The values for τ_1 and τ_2 are $(\sqrt{2N})^{-1}$ and $(\sqrt{2\sqrt{N}})^{-1}$, respectively, as suggested in [39], where $2N$ is the dimension of the regularization strategy $\mathcal{S}_p$.

For each regularization strategy $\mathcal{S}_p$ in the population, we can in principle use it to restore a blurred image, and then build up the histogram $\hat{p}_\kappa(\kappa)$ by evaluating κ at each pixel and counting their relative occurrence frequencies. Finally, we can evaluate the ETC-pdf error $E^\kappa_{pdf}(\mathcal{S}_p)$ as a function of the strategy $\mathcal{S}_p$ by comparing the difference of the normalized histogram and the model ETC-pdf, and use this value as our basis for competition and selection. For the selection stage, we adopt tournament competition as our selection mechanism, where we generate a subset $T(\mathcal{S}_p) \subset P \setminus \{\mathcal{S}_p\}$ for each strategy $\mathcal{S}_p$ with $|T| = Q$ by randomly sampling the population Q times. We can then form the set $\mathcal{W}(\mathcal{S}_p)$ as below

$$\mathcal{W}(\mathcal{S}_p) = \{\mathcal{S}_q \in T : E^\kappa_{pdf}(\mathcal{S}_q) > E^\kappa_{pdf}(\mathcal{S}_p)\} \qquad (6.23)$$

which contains all those strategies $\mathcal{S}_q$ in $\mathcal{T}$ with their corresponding ETC-pdf error greater than that of $\mathcal{S}_p$. We then define the *win count* $w_c(\mathcal{S}_p)$ as the cardinality of $\mathcal{W}(\mathcal{S}_p)$

$$w_c(\mathcal{S}_p) = |\mathcal{W}(\mathcal{S}_p)| \tag{6.24}$$

Denoting the population at the current generation as $P(t)$, we order the regularization strategies $\mathcal{S}_p$ in $P(t)$ according to decreasing values of their associated win counts $w_c(\mathcal{S}_p)$, and choose the first μ individuals in the list to be incorporated into the next generation $P(t+1)$.

6.4.1   Competition under Approximate Fitness Criterion

As we have mentioned, we can in principle perform a restoration for each of the 2μ regularization strategies in $P(t)$ to evaluate its fitness, but each restoration is costly in terms of the amount of computations, since it has to be implemented iteratively, and is especially the case when the number of individuals in the population is large. We therefore resort to an approximate competition and selection process where each individual strategy in the population is used to restore only a part of the image. Specifically, we associate each strategy $\mathcal{S}_p, p = 1, \ldots, 2\mu$ in the population $P(t)$ with a subset $\mathcal{R}_p \subset \mathcal{X}$, where $\mathcal{X}$ denotes the set of points in an $N_y \times N_x$ image lattice

$$\mathcal{X} = \{(i_1, i_2) : 1 \le i_1 \le N_y, 1 \le i_2 \le N_x\} \tag{6.25}$$

and where the regions $\mathcal{R}_p$ form a partition of $\mathcal{X}$

$$\mathcal{R}_{p_1} \cap \mathcal{R}_{p_2} = \emptyset \qquad p_1 \ne p_2 \tag{6.26}$$

$$\bigcup_{p=1}^{2\mu} \mathcal{R}_p = \mathcal{X} \tag{6.27}$$

In this way, we can define the quantity $\hat{E}_{pdf}^{\kappa}(\mathcal{S}_p, \mathcal{R}_p)$, which is both a function of $\mathcal{S}_p$ and $\mathcal{R}_p$, as the evaluation of $E_{pdf}^{\kappa}(\mathcal{S}_p)$ restricted to the subset $\mathcal{R}_p$ only. This serves as an estimate of the exact $E_{pdf}^{\kappa}(\mathcal{S}_p)$, which is evaluated over the entire image lattice $\mathcal{X}$ and is therefore a function of $\mathcal{S}_p$ only. In order to approximate the original error function closely, we must choose the subsets $\mathcal{R}_p$ in such a way that each of them captures the essential features of the distribution of the ETC measure in the complete image. This implies that each of these subsets should be composed of disjoint regions scattered almost uniformly throughout the lattice $\mathcal{X}$ and that both the ways of aggregating these regions into a single $\mathcal{R}_p$ and the ways of assigning these subsets to each individual in the population should be randomized in each generation. In view of these, we partition the image lattice $\mathcal{X}$ into non-overlapping square blocks B_s of size $n \times n$

$$\mathcal{B} = \{B_s : s \in \mathcal{I}\} \tag{6.28}$$

where

$$B_{s_1} \cap B_{s_2} = \emptyset \qquad s_1 \neq s_2 \tag{6.29}$$

$$\bigcup_{s \in \mathcal{I}} B_s = \mathcal{X} \tag{6.30}$$

In this equation, $\mathcal{I}$ is the index set $\{1, \ldots, S\}$ and S is the number of $n \times n$ square blocks in the image lattice. Assuming $2\mu < S$ and furthermore that $u \equiv S/2\mu$ is an integer, i.e., 2μ is a factor of S, we can construct the following *random* partition $\mathcal{P}_{\mathcal{I}} = \{\mathcal{I}_p : p = 1, \ldots, 2\mu\}$ of $\mathcal{I}$ where

$$\mathcal{I}_{p_1} \cap \mathcal{I}_{p_2} = \emptyset \qquad p_1 \neq p_2 \tag{6.31}$$

$$\bigcup_{p=1}^{2\mu} \mathcal{I}_p = \mathcal{I} \tag{6.32}$$

and $|\mathcal{I}_p| = u$ as follows:

- Randomly sample the index set $\mathcal{I}$ u times without replacement to form $\mathcal{I}_1$.

- For $p = 2, \ldots, 2\mu$, randomly sample the set $\mathcal{I} \setminus \bigcup_{q=1}^{p-1} \mathcal{I}_q$ u times to form $\mathcal{I}_p$

Finally, we define the subsets $\mathcal{R}_p$ corresponding to each individual strategy $\mathcal{S}_p$ as follows.

$$\mathcal{R}_p = \bigcup_{s \in \mathcal{I}_p} B_s \tag{6.33}$$

We can then evaluate $\hat{E}_{pdf}^\kappa(\mathcal{S}_p, \mathcal{R}_p)$ for each strategy $\mathcal{S}_p$ with its corresponding region $\mathcal{R}_p$ and use it in place of the exact error $E_{pdf}^\kappa(\mathcal{S}_p)$ for the purpose of carrying out the tournament competition and ranking operation in Section 6.4. The process is illustrated in Figure 6.5. In this case, for the purpose of illustration, we have used a population with only four individuals. The image is divided into 12 sub-blocks and we have assigned 3 sub-blocks to each individual in the population. This assignment is valid only for one update iteration only, and a randomly generated alternative assignment is adopted for the next iteration to ensure adequate representation of the entire image through these assigned blocks.

6.4.2 Choice of Optimal Regularization Strategy

After the tournament competition and ranking operations, we can in principle choose the optimal strategy $\mathcal{S}^*$ from the population $P(t)$ and use it to restore the image for a single iteration. We can define optimality here in several ways. The obvious definition is to choose that strategy $\mathcal{S}^*$ with the minimum value of $\hat{E}_{pdf}^\kappa$

$$\mathcal{S}^* = \arg\min_p \hat{E}_{pdf}^\kappa(\mathcal{S}_p, \mathcal{R}_p) \tag{6.34}$$

1	4	3	2
3	4	1	3
4	1	2	2

Figure 6.5: Illustrating the assignment of regions $\mathcal{R}_p$ to individual regularization strategy $\mathcal{S}_p$ for a 4-member population

Alternatively, we can form a subset of those elite strategies with their win index $w_c(\mathcal{S}_p)$ equal to the maximum possible value, i.e., the tournament size Q:

$$\mathcal{E}_Q = \{\mathcal{S}_p : w_c(\mathcal{S}_p) = Q\} \subset P(t) \tag{6.35}$$

and then choose our strategy $\mathcal{S}^*$ by uniformly sampling from this subset.

The above selection schemes are suitable for the competition and selection stage of conventional evolutionary computational algorithms where the fitness value of each individual directly reflects its inherent optimality. In the current approximate competition scheme; however, we have replaced the exact fitness value E_{pdf}^κ with the estimate $\hat{E}_{pdf}^\kappa$, which depends on both the inherent fitness of the strategy $\mathcal{S}_p$ and its particular assigned region $\mathcal{R}_p$. Therefore, there may exist cases where a non-optimal strategy will acquire a high fitness score due to a chance combination of image blocks B_s in forming its assigned region. To prevent this, a more sophisticated selection scheme involving the elite individuals in the population is required.

Recall that the estimated error $\hat{E}_{pdf}^\kappa$ is both a function of the inherent fitness of the strategy $\mathcal{S}_p$ and its assigned region $\mathcal{R}_p$, and a particular combination of image blocks B_s may result in a low estimated error even though the inherent fitness of $\mathcal{S}_p$ is not very high. However, we should expect that, for a strategy with low inherent fitness, it will quickly encounter a combination of image blocks B_s leading to a very large estimated error and become displaced from the population. On the other hand, for a strategy with high inherent fitness, we should expect that the corresponding estimated error will be low for a variety of image block combinations in each generation, and it will most probably survive into the next generation. In view of this, a proper way to estimate an optimal regularization strategy from the current population should involve the *survival time* t_p of each individual strategy $\mathcal{S}_p$ which is defined as follows:

$$t_p \equiv t - t_p^i \tag{6.36}$$

such that

$$\mathcal{S}_p \in \bigcap_{t'=t_p^i}^{t} P(t') \tag{6.37}$$

In other words, t_p^i is the generation where the strategy $\mathcal{S}_p$ first appears in the population, and t is the current generation. The survival time t_p is defined as the difference between these two indices. In general, it is reasonable to assume that a regularization strategy with a long survival time t_p is more likely to possess high inherent fitness, but whether a strategy with a short survival time possesses high inherent fitness is yet to be confirmed in later generations. As a result, it is more reasonable to choose the optimal regularization strategy based on those in the population with long survival times. Rearranging the corresponding survival time t_p of each individual in ascending order,

$$t_{(1)}, \ldots, t_{(p)}, \ldots, t_{(2\mu)} \tag{6.38}$$

such that

$$t_{(1)} \le \cdots \le t_{(p)} \le \cdots \le t_{(2\mu)} \tag{6.39}$$

and $t_{(p)}$ denotes the pth order statistic of the survival time sequence, we expect those values $t_{(p)}$ with large index p will most likely correspond to a strategy $\mathcal{S}_{(p)}$ with high inherent fitness. Choosing $p_0 > 1$ and regarding each strategy as a vector in $\mathbf{R}^8$, we define the following *combined regularization strategy*

$$\mathcal{S}_{p_0}^* = \frac{1}{2\mu - p_0 + 1} \sum_{p=p_0}^{2\mu} \mathcal{S}_{(p)} \tag{6.40}$$

To ensure the inclusion of only those individuals with high inherent fitness in the above averaging operation, the value p_0 is usually chosen such that $p_0 \gg 1$.

In addition, if we possess *a priori* knowledge regarding inherent properties of desirable regularization strategies characterized by the *constraint set* $\mathcal{C}_\mathcal{S}$, we can modify the previous averaging procedure to include this knowledge as follows

$$\mathcal{S}_{p_0}^* = \frac{\sum_{p=p_0}^{2\mu} I_{\{\mathcal{S}_{(p)} \in \mathcal{C}_\mathcal{S}\}} \mathcal{S}_{(p)}}{\sum_{p=p_0}^{2\mu} I_{\{\mathcal{S}_{(p)} \in \mathcal{C}_\mathcal{S}\}}} \tag{6.41}$$

where $I_{\{\mathcal{S}_{(p)} \in \mathcal{C}_\mathcal{S}\}}$ is the indicator function of $\mathcal{C}_\mathcal{S}$.

The optimal strategy $\mathcal{S}_{p_0}^*$ is constructed based on the estimated ETC-pdf error $\hat{E}_{pdf}^\kappa$, and we have to evaluate its performance based on the true error measure E_{pdf}^κ eventually, which requires using $\mathcal{S}_{p_0}^*$ to restore the whole image for one iteration and then collecting the statistics of the measure κ on the resulting image to form an estimation of $p_\kappa(\kappa)$. In the current algorithm, this has to be performed once only, again highlighting the advantage of using $\hat{E}_{pdf}^\kappa$, which requires performing restoration over an image subset only for each individual strategy in the population, instead of using E_{pdf}^κ in the competition process, which requires performing restoration for an entire image. In this way, the time required for fitness evaluation is independent of the number of individuals in the population.

Denoting the current optimal strategy as $\mathcal{S}^*(t)$ and the last strategy used to update the image as $\mathcal{S}^*(t-1)$, we should decide between using $\mathcal{S}^*_{p_0}$ or $\mathcal{S}^*(t-1)$ as $\mathcal{S}^*(t)$, or we should simply left the image unrestored for the present iteration. Denoting the *null strategy* as $\mathcal{S}_\phi$, which amounts to leaving the present image unrestored, the basis of this decision should be the true ETC-pdf error value $E^\kappa_{pdf}(\mathcal{S})$ where $\mathcal{S} = \mathcal{S}^*_{p_0}, \mathcal{S}^*(t-1)$ or $\mathcal{S}_\phi$. Using $\hat{\mathbf{f}}_r(\mathcal{S})$ to indicate the resulting restored image through the action of $\mathcal{S}$, $\hat{\mathbf{f}}(t)$ as the updated image, and $\hat{\mathbf{f}}(t-1)$ as the pre-updated image, we adopt the following decision rule for choosing $\mathcal{S}^*(t)$

- If $E^\kappa_{pdf}(\mathcal{S}^*_{p_0}) = \min\{E^\kappa_{pdf}(\mathcal{S}^*_{p_0}), E^\kappa_{pdf}(\mathcal{S}^*(t-1)), E^\kappa_{pdf}(\mathcal{S}_\phi)\}$

  1. $\mathcal{S}^*(t) = \mathcal{S}^*_{p_0}$
  2. $\hat{\mathbf{f}}(t) = \hat{\mathbf{f}}_r(\mathcal{S}^*_{p_0})$

- If $E^\kappa_{pdf}(\mathcal{S}^*(t-1)) = \min\{(E^\kappa_{pdf}(\mathcal{S}^*_{p_0}), E^\kappa_{pdf}(\mathcal{S}^*(t-1)), E^\kappa_{pdf}(\mathcal{S}_\phi)\}$

  1. $\mathcal{S}^*(t) = \mathcal{S}^*(t-1)$
  2. $\hat{\mathbf{f}}(t) = \hat{\mathbf{f}}_r(\mathcal{S}^*(t-1))$

- If $E^\kappa_{pdf}(\mathcal{S}_\phi) = \min\{(E^\kappa_{pdf}(\mathcal{S}^*_{p_0}), E^\kappa_{pdf}(\mathcal{S}^*(t-1)), E^\kappa_{pdf}(\mathcal{S}_\phi)\}$

  1. $\mathcal{S}^*(t) = \mathcal{S}_\phi$
  2. $\hat{\mathbf{f}}(t) = \hat{\mathbf{f}}_r(\mathcal{S}_\phi) = \mathbf{x}(t-1)$

6.5   Experimental Results

The current algorithm was applied to the same set of images in the Chapter 5. The parameters of the ETC-pdf image model were chosen as follows: $a_1 = 0.83, a_2 = 0.85$ and $\kappa_3 = 1.8$ using a 5×5 averaging window. The value $\triangle = 0.1$ is adopted as the bin width. For the EP algorithm we have chosen $\mu = 16$, a sub-block size of 32×32 for B_s, and a tournament size of $Q = 10$ in the selection stage.

In Figure 6.6, we apply the algorithm to the flower image. The degraded image is shown in Figure 6.6a, and the restored image using the EP algorithm is shown in Figure 6.6d. For comparison purposes, we also include the same image restored using alternative algorithms in Figures 6.6b to c. Figure 6.6b shows the restored image using the non-adaptive approach, where a single $\lambda = 1/\text{BSNR}$ is adopted for the whole image. We can notice the noisy appearance of the resulting restoration when compared with Figure 6.6d. We have also included the HMBNN restoration results in Figure 6.6c, and we can notice that the qualities of the two restored images are comparable.

In Figure 6.7, we apply the algorithm to the flower image degraded by a 5×5 uniform PSF at an increased noise level of 20 dB BSNR. The degraded image is shown in Figure 6.7a. The increased noise level manifests itself in the more

(a)   (b)

(c)   (d)

Figure 6.6: Restoration of the flower image (5×5 uniform blur, 30 dB BSNR). (a) Blurred image, (b) non-adaptive approach ($\lambda = 1/\text{BSNR}$), (c) HMBNN, (d) EP

Figure 6.7: Restoration of the flower image (5×5 uniform blur, 20 dB BSNR). (a) Blurred image, (b) non-adaptive approach ($\lambda = 1/BSNR$), (c) HMBNN, (d)EP

Figure 6.8: Restoration of the Lena image (5×5 uniform blur). (a)-(b) Degraded images: (a) 30 dB BSNR, (b) 20 dB BSNR, (c)-(d) restoration results using EP: (c) 30 dB BSNR, (d) 20 dB BSNR

noisy appearance of the non-adaptive restoration result in Figure 6.7b. This can be compared with our result in Figure 6.7d, where the adoption of the ETC-pdf error criterion allows the re-determination of the relevant parameters through EP. We have also included the HMBNN result in Figure 6.7c. Again, the two restored results are comparable, although we can notice a slightly more noisy appearance in the EP result.

We also apply the EP restoration algorithm to the image in Figure 6.8. Figures 6.8a and b show the degraded images using a 5×5 uniform PSF with 30 dB and 20 dB noise added respectively and the EP restoration results in Figures 6.8c and d.

The λ assignment maps for the flower image under 5×5 uniform blur are shown in Figure 6.9. Figure 6.9a shows the assignment map under 30 dB additive

(a)   (b)

Figure 6.9: Distribution of λ values for the flower image: (a) 30 dB BSNR, (b) 20 dB BSNR

noise, and Figure 6.9b shows the corresponding map under 20 dB noise. In the two maps, the darker gray values correspond to small λ values, and the brighter values correspond to large λ values. In general, the regularization strategy discovered by the artificial evolutionary process assigns small parameter values to edge/textured regions. These smaller values in turn help to bring out more fine details in these regions in the accompanying restoration phase. On the other hand, large λ values are assigned to the smooth regions for noise suppression. In addition, the maps for the same image are different under different levels of additive noise: at low noise levels, the area over which large λ values are assigned is small compared with the corresponding maps under high noise levels. This implies that edge/texture enhancement takes precedence over noise suppression at low noise levels. On the other hand, for higher noise levels, most of the areas are assigned large λ values, and only the very strong edges and textured regions are assigned moderately smaller λ values. We can thus conclude that at low noise levels, the primary purpose is edge/texture enhancement, whereas for higher noise levels it is noise removal.

This can also be confirmed from Figure 6.10, where we show plots of the regularization profile corresponding to the λ-maps in Figure 6.10 for the flower image (the logarithmic plots are shown due to the large dynamic ranges of the parameters). It can be seen that the profile corresponding to 20 dB noise is shifted to the right with respect to the 30 dB profile, which implies a larger value of the threshold parameter α_p and resulting in a larger image area being classified as smooth regions. This agrees with our previous conclusion that noise suppression takes precedence over edge/texture enhancement when the noise level is high.

It is also seen from the figure that pixels with high local standard deviation, which possibly correspond to significant image features, are assigned different

Figure 6.10: Regularization profiles under different noise levels for the flower image

(a)   (b)

Figure 6.11: Further restoration results (5×5 uniform blur, 30 dB BSNR). (a) Blurred eagle image, (b) restored image using EP

values of λ_i at different noise levels. At lower noise levels, the λ_i values assigned at large σ_i are seen to be smaller than the corresponding values at higher noise levels. This is reasonable due to the possibility of excessive noise amplification at higher noise levels, which in turn requires higher values of λ_i for additional noise suppression.

We have also applied this algorithm to the eagle image. Figure 6.11a shows the degraded images for eagle under 5×5 uniform blur at 30 dB BSNR and Figure 6.11b shows the result using the current EP approach. We can more readily appreciate the importance of adaptive processing from these additional results.

We list the RMSE of the restored images using the current algorithm, together with those restored using the HMBNN algorithm, in Table 6.1. Each RMSE value for the EP algorithm is evaluated as the mean of five independent runs (the values in the parentheses indicate the standard deviation among the different runs). It is seen that, in all the cases, the mean RMSE values of the EP algorithm are comparable to the HMBNN error values.

We also include the standard deviation associated with each RMSE value for the EP results over five independent runs. We can see that the values of these standard variations are within acceptable ranges. It was also observed that there are no noticeable differences between the appearances of the restored images in different trials. As a result, we can conclude that these variations are acceptable for the current problem.

Table 6.1: RMSE values of the restoration results using different restoration algorithms

Image, noise level	Blurred	Non-adaptive	HMBNN	EP
Lena, 30 dB	13.03	8.65	7.09	7.35(0.0611)
Lena, 20 dB	13.79	10.56	9.69	9.78(0.112)
Eagle, 30 dB	9.88	10.28	6.89	7.13(0.0600)
Eagle, 20 dB	11.98	11.57	8.99	8.97(0.0364)
Flower, 30 dB	8.73	7.70	4.98	5.45(0.0310)
Flower, 20 dB	10.06	8.47	6.92	7.06(0.0767)

6.6 Other Evolutionary Approaches for Image Restoration

An alternative evolutionary computational approach to adaptive image regularization based on a neural network model, the Hierarchical Cluster Model (HCM) is proposed in [144]. An HCM is a hierarchical neural network that coordinates parallel, distributed sub-networks in a recursive fashion with its cluster structures closely matching the homogeneous regions of an image. In addition, its sparse synaptic connections are effective in reducing the computational cost of restoration. In the proposed algorithm, the degraded image is first segmented into different regions based on its local statistics. They are further partitioned into separate clusters and mapped onto a three-level HCM. Each image cluster, equipped with an optimized regularization parameter λ, performs restoration using the model-based neuron updating rules.

The main difference between this algorithm and our previous approaches is that an evolutionary strategy scheme is adopted to optimize the λ values by minimizing the HCM energy function. Specifically, the scheme progressively selects the well-evolved individuals that consists of a set of partially restored images with their associated cluster structures, segmentation maps and the optimized λ values. Once the restoration is completed, the final empirical relationship between the optimized λ values and a specific local perception measure can be determined and reused in the restoration of other unseen, degraded images, which will remove the computational overhead of evolutionary optimization. The specific features of this approach are described below.

6.6.1 Hierarchical Cluster Model

Hierarchical Cluster Model (HCM) is a nested neural network that consists of parallel, distributed sub-networks or clusters. It models the organization of the neocortex in human brain where functional groups of neurons organize themselves dynamically into multidimensional subnetworks. HCM can be constructed in a bottom-up manner. Individual neurons form the trivial level 0 clusters of the network. The neurons sharing similar functionality and characteristics coalesce into a number of level 1 clusters. Likewise, those level 1 clusters with homogeneous characteristics in turn coalesce into level 2 clusters. This process continues repeatedly to form a hierarchical neural network with multidimensional distributed clusters.

A three-level HCM is employed in the context of image regularization. There exists a close correspondence between the chosen HCM with image formation. The individual neuron or level 0 cluster of HCM corresponds to each image pixel. Similarly, level 1 clusters correspond to homogeneous image regions whereas level 2 clusters correspond to dynamic boundary adjustments between regions. The algorithm is composed of the following two stages.

6.6.2   Image Segmentation and Cluster Formation

In this first stage, a degraded image is segmented into different homogeneous clusters and restored with adaptive λ values to achieve optimum visual result. A three-level HCM is constructed from an image in two steps: an image is first segmented into different regions based on its local statistics and each individual region is then further partitioned into a set of clusters.

6.6.3   Evolutionary Strategy Optimization

A (μ, τ, ν, ρ) evolutionary scheme is adopted to optimize the regularization parameter values. The scheme uses μ parents, ν offspring with τ as the upper limit of life span, and ρ as the number of ancestors for each descendant. Since the regularization parameter associated with each cluster is closely related to the perception measure such as cluster average local variance, the evolutionary scheme continually explores the optimum relationship between the local λ value and the cluster average local variance by minimizing a suitable error measure. In this algorithm, a parameterized logarithmic function is adopted as a model for the non-linear transformation in each cluster with the associated parameters of the function determined through the evolutionary scheme. In addition, the energy function of the HCM is adopted as the fitness function, the value of which is optimized through the two operations of mutation and tournament selection.

As a result, compared with the current algorithm, this alternative evolutionary restoration algorithm offers the possibility of refined local image characterization by adopting multiple sets of regularization profiles at the expense of increased computational requirements (due to the need to adopt EC to search for the optimal profile for each region), and the necessity to perform a prior segmentation of the image.

6.7   Summary

We have proposed an alternative solution to the problem of adaptive regularization in image restoration in the form of an artificial evolutionary algorithm. We first characterize an image by a model discrete probability density function (pdf) of the ETC measure κ, which reflects the degree of correlation around each image pixel, and effectively characterizes smooth regions, textures and edges. An optimally regularized image is thus defined as the one with its corresponding ETC-pdf closest to this model ETC-pdf. In other words, during the restoration process, we have to minimize the difference between the ETC-pdf of the restored image, which is usually approximated by the ETC histogram, and the model ETC-pdf. The discrete nature of the ETC histogram and the non-differentiability of the resulting cost function necessitates the use of evolutionary programming (EP) as our optimization algorithm to minimize the error function. The population-based approach of evolutionary programming provides an efficient method to search for potential optimizers of highly irregular and non-differentiable cost function as the current *ETC-pdf error measure*. In addition, the current problem is also non-stationary, as the optimal regularization strategy in the current iteration of pixel updates is not necessarily the same as the next iteration. The maintenance of a diversity of potential optimizers in the evolutionary approach increases the probability of finding alternative optimizers for the changing cost function. Most significantly, the very adoption of evolutionary programming has allowed us to broaden the range of cost functions in image processing which may be more relevant to the current application, instead of being restricted to differentiable cost functions.

Chapter 7

Blind Image Deconvolution

7.1 Introduction

We will again restrict the presentation of this chapter to the popular linear image degradation model:

$$\mathbf{g} = \mathbf{H}\hat{\mathbf{f}} + \mathbf{n} \tag{7.1}$$

where $\mathbf{g}$, $\mathbf{f}$ and $\mathbf{n}$ are the lexicographically ordered degraded image, original image and additive white Gaussian noise (AWGN), respectively [21, 145]. $\mathbf{H}$ is the linear distortion operator determined by the point spread function (PSF), $\mathbf{h}$. Blind image deconvolution is an inverse problem of rendering the best estimates, $\hat{\mathbf{f}}$ and $\hat{\mathbf{h}}$, to the original image and the blur based on the degradation model. It is a difficult ill-posed problem as the uniqueness and stability of the solution is not guaranteed [24]. Classical restorations require complete knowledge of the blur to be known prior to restoration [21, 23, 145] as discussed in the previous chapters. However, it is often too costly, cumbersome or, in some cases, impossible to determine the exact blur *a priori*. These could be due to various practical constraints such as the difficulty of characterizing air turbulence in aerial imaging, or the potential health hazard of employing a stronger incident beam to improve the image quality in X-ray imaging. In these circumstances, blind image deconvolutions are essential in recovering visual clarity from the degraded images.

Traditional blind methods such as *a priori* blur identifications formulate blind restoration into two disjoint processes where the blur is first estimated, followed by classical restoration based on the identified blur [146, 147]. The methods are inflexible as they require the parametric structures of the blur to be known exactly, and are tailored specifically for the targeted blur type. In addition, they are ineffectual in identifying certain blurs such as a Gaussian mask, which does not exhibit prominent frequency nulls.

The success of linear regression theory in digital communication and signal processing motivates various researchers to consider a degraded image as an autoregressive moving average (ARMA) process. The original image is modeled as an autoregressive (AR) process, with the blur representing the moving average (MA) coefficients. Under this assumption, blind image deconvolution is transformed into an ARMA parameter estimation problem. Maximum likelihood (ML) [148] and generalized cross validation (GCV) [149] are two popular techniques used to perform the estimation. ML is commonly used in conjunction with expectation maximization (EM) to determine the parameters by maximizing their log-likelihood probabilities from the solution space. The side effect is that it suffers from insensitivity toward changes in individual parameters due to its huge input argument list. GCV evaluates the parameters by minimizing the weighted sum of the predictive errors. Its main disadvantage lies in the repetitive process of cross validations, which constitutes significant computational cost. Both methods require small AR and MA support dimensions, albeit at the cost of diminished modeling effectiveness. The problem is further complicated by the ARMA image stationarity constraint that is inconsistent with some real-life images consisting of inhomogeneous smooth, textured, and edge regions.

The conceptual simplicity of iterative image and blur estimations has attracted researchers to propose various iterative blind deconvolution schemes. These include iterative blind deconvolution (IBD) [150], simulated annealing (SA) [151], and non-negativity and support constraints recursive inverse filtering (NAS-RIF) [152]. IBD alternates between spatial and frequency domains, imposing constraints onto the image and blur estimates repeatedly. In contrast, SA employs the standard annealing procedure to minimize a multimodal cost function. NAS-RIF extends the recursive filtering of the blurred image to optimize a convex cost function. These iterative algorithms require the image objects to have known support dimensions, and to be located in a uniform background, and thus impose restrictions on many applications. Other difficulties such as sensitivity to initial conditions, slow convergence, and interdomain dependency further complicate the iterative schemes.

Recent investigations into image regularization have been extended to address blind image deconvolution [153, 154]. The problem is formulated into two symmetrical processes of image restoration and blur identification. However, the implicit symmetrical assumption conflicts with several observations. First, most blur functions have predominantly low-frequency contents. In contrast, the frequency spectrums of typical images vary from low in the smooth regions, to medium and high in the textured and edge regions. Second, studies show that practical blurs satisfy up to a certain degree of parametric structure. In comparison, we have little or no prior knowledge about the imaging scenes. Third, the methods ignore different characteristics of the image and blur domains, leading to poor incorporation of their unique properties. Recent work alleviates these difficulties by integrating the unique properties of image and blur domains into a recursive scheme based on soft blur identification and a hierarchical neural network [155]. It assigns different emphases to image restoration and blur identification according to their characteristics, thereby providing a priority-based subspace deconvolution.

This chapter presents two recent approaches to adaptive blind image deconvolution. The first is based on computational reinforced learning in an attractor-embedded solution space, and and the second is a recursive scheme based on soft-decision blur identification and a hierarchical neural network.

7.1.1 Computational Reinforced Learning

The computational reinforced learning (CRL) method, an extended evolutionary strategy that integrates the merits of priority-based subspace deconvolution, is developed to generate the improved blurs and images progressively. Evolutionary algorithms (EA) are often used to solve difficult optimization problems where traditional path-oriented and volume-oriented methods fail [156, 157]. Evolutionary strategies (ES), a mainstream of EA, model the evolutionary principles at the level of individuals or phenotypes. An important feature of this method lies in its capability to alleviate the common difficulties encountered by other blind algorithms, namely, interdomain dependency, poor convergence, and local minima trapping. It allows the most tangible performance measure to be adopted, rather than being restricted by the criteria's tractability as in other schemes. We embrace this feature by introducing a novel entropy-based performance measure that is both effective and intuitive in assessing the fitness of the solutions, as well as estimating the blur support dimensionality.

As blind deconvolution exhibits the property of reducibility, careful incorporation of image and blur information are instrumental in achieving good results. In accordance with the notion, a mutation attractor space is constructed by incorporating the blur knowledge domain into the algorithm. A maximum *a posteriori* estimator is developed to predict these attractors, and their relevance is evaluated within the evolutionary framework. A novel reinforced mutation scheme that integrates classical mutation and reinforced learning is derived. The technique combines self-adaptation of the attractors and intelligent reinforced learning to improve the algorithmic convergence, thereby reducing the computational cost significantly. The new evolutionary scheme features a multithreaded restoration that is robust toward divergence and poor local minima trappings.

Unlike traditional blind methods where a hard decision on the blur structure has to be made prior to restoration, this approach offers a continual relevance-feedback learning throughout image deconvolution. This addresses the formulation dilemma encountered by other methods, namely, integrating the information of well-known blur structures without compromising its overall flexibility. As image restoration is a high-dimensional problem, fast convergence is particularly desirable. Therefore, a stochastic initialization procedure is introduced to speed the initial searching. A probability neural network is devised to perform landscape surveying and blur dimensionality estimation. The underlying principle is that the solution subspace with better performance should, accordingly, be given more emphasis.

7.1.2 Blur Identification by Recursive Soft Decision

Similar to the CRL method, the soft-decision method offers a continual soft-decision blur adaptation throughout the restoration. It incorporates the knowledge of standard blur structure while preserving the flexibility of blind restoration scheme. A new cost function that comprises the data fidelity measure, image and blur-domain regularization terms, and a novel soft-decision blur estimation error is introduced. A blur estimator is devised to determine various soft parametric blur estimates. The significance of the estimates is evaluated by its proximity to the currently computed blur.

In this method, we formulate blind image restoration into a two-stage subspace optimization. The overall cost function is projected and optimized with respect to the image and blur domains recursively. A nested neural network, called Hierarchical Cluster Model (HCM) [114, 158] is employed to provide an adaptive, perception-based restoration by minimizing the image-domain cost function [159]. Its sparse synaptic connections are effective in reducing the computational cost of restoration. A blur compensation scheme is developed to rectify the statistical averaging effect that arises from the ambiguous blur identifications in the edge and texture regions. This method does not require assumptions such as image stationarity as in ARMA modeling, or known support objects in uniform backgrounds as in deterministic constraint restoration. The approach also addresses the asymmetry between image restoration and blur identification by employing the HCM to provide a perception-based restoration and conjugate gradient optimization to identify the blur.

Conjugate gradient optimization is adopted, wherever appropriate, in image restoration and blur identification. It is chosen ahead of other gradient-based approaches such as Quasi-Newton and steepest descent techniques as it offers a good compromise in terms of robustness, low computational complexity and small storage requirement.

7.2   Computational Reinforced Learning

7.2.1   Formulation of Blind Image Deconvolution as an Evolutionary Strategy

Classical image restorations usually involve the minimization of a quadratic cost function:

$$J(\hat{\mathbf{f}}) = \frac{1}{2}\,\hat{\mathbf{f}}^T\mathbf{Q}\hat{\mathbf{f}} + \mathbf{r}^T\hat{\mathbf{f}} \qquad (7.2)$$

As the matrix $\mathbf{Q}$ is symmetric positive definite, the minimization of $J(\hat{\mathbf{f}})$ will lead to the restored image, $\hat{\mathbf{f}}$. In comparison, blind image deconvolution is an inverse problem of inferring the best estimates, $\hat{\mathbf{f}}$ and $\hat{\mathbf{h}}$, to the original image and the blur. Its formulation involves the development and optimization of a multimodal cost function $J(\hat{\mathbf{f}},\hat{\mathbf{h}}|\mathbf{g})$.

As the image and blur domains exhibit distinctive spatial and frequency profiles, we project and minimize the cost function $J(\hat{\mathbf{f}},\hat{\mathbf{h}}|\mathbf{g})$ with respect to each domain iteratively. The scheme can be summarized by:

i)   Initialize $\hat{\mathbf{h}}$

ii) Recursively minimize the ith iterative subspace cost functions:

$$\min J_i(\hat{\mathbf{f}}|\mathbf{g},\hat{\mathbf{h}}) = \min \mathbf{P}_{\hat{\mathbf{f}}}\left\{J_i(\hat{\mathbf{f}},\hat{\mathbf{h}}|\mathbf{g})\right\} \tag{7.3}$$

and

$$\min J_i(\hat{\mathbf{h}}|\mathbf{g},\hat{\mathbf{f}}) = \min \mathbf{P}_{\hat{\mathbf{h}}}\left\{J_i(\hat{\mathbf{f}},\hat{\mathbf{h}}|\mathbf{g})\right\} \tag{7.4}$$

iii) Stop when the termination criterion is satisfied

$J_i(\hat{\mathbf{f}}|\mathbf{g},\hat{\mathbf{h}})$ and $J_i(\hat{\mathbf{h}}|\mathbf{g},\hat{\mathbf{f}})$ are the ith iterative subspace cost functions, and $\mathbf{P}_{\hat{\mathbf{f}}}$ and $\mathbf{P}_{\hat{\mathbf{h}}}$ are the projection operators with respect to the image and blur domains. The advantages of these approaches include the simplicity of constraint and algorithmic formulation. The cost function $J(\hat{\mathbf{f}},\hat{\mathbf{h}}|\mathbf{g})$ is usually chosen to be quadratic with respect to the image and blur domains, thereby ensuring convergence in their respective subspaces. However, the iterative projections create interdependency between $\hat{\mathbf{f}}$ and $\hat{\mathbf{h}}$, giving rise to potential poor convergence and local minima trapping. This often leads to inadequate image restorations and blur identifications.

To address these difficulties, we extend the alternating minimization procedure into an (μ, κ, ν, ρ) evolutionary strategy. The scheme uses μ parents, ν offspring with κ as the upper limit of life span, and ρ as the number of ancestors for each descendant. The mathematical representation of the new scheme can be summarized by:

i) Initialize Φ_0

ii) For ith generation, determine the dynamic image and blur solution spaces, Ω_i and Φ_i:

$$\Omega_i = \left\{\hat{\mathbf{f}}_i|\hat{\mathbf{f}}_i = \arg\min_{\hat{\mathbf{f}}} J(\hat{\mathbf{f}}|\mathbf{g},\hat{\mathbf{h}}_i \in \Phi_i)\right\} \tag{7.5}$$

$$\Phi_i = \left\{\hat{\mathbf{h}}_i|\hat{\mathbf{h}}_i = \mathbf{M} \circ \mathbf{R} \circ \mathbf{S} \circ \mathbf{F}(\hat{\mathbf{f}}_{i-1}|\hat{\mathbf{f}}_{i-1} \in \Omega_{i-1})\right\} \tag{7.6}$$

iii) Stop when convergence or the maximum number of generation is reached

The blur solution space Φ_i is generated based on concatenation of the performance evaluation operator $\mathbf{F}$, the candidate selection operator $\mathbf{S}$, the niche-space recombination operator $\mathbf{R}$ and the reinforced mutation operator $\mathbf{M}$. The new technique preserves the algorithmic simplicity of the projection-based deconvolutions by performing image restoration in (7.5) and blur identification in (7.6). Moreover, it exploits the virtue of the evolutionary strategies to alleviate inter-domain dependency and poor convergence, thereby enhancing the robustness of the deconvolution scheme.

Probability-Based Stochastic Initialization

The main challenge of applying evolutionary algorithms lies in the significant computational cost associated usually with the evolution of good solutions and characteristics. In the context of blind image deconvolution, the inherent high dimensionality puts further demands on computational efficiency. One of the key issues in blind image deconvolution is the determination of blur dimensionality. Most blind methods either assume the blur supports are known *a priori*, or impose some highly restrictive assumptions to estimate them. The combination of high image dimensionality and uncertainty in blur supports indicates the importance of intelligent initialization. In view of this, we introduce a new probability neural network shown in Figure 7.1 to provide a performance-feedback initialization.

The network consists of three layers, namely, input, intermediary and output layers. The input layer comprises neurons representing the blur knowledge base $\mathbf{x}_i, i = 1, \ldots, K$, and a single neuron representing the stochastic perturbation, $\mathbf{x}_s$. The knowledge base consists of K established blur structures that can be adjusted to incorporate any information or previous experience. The intermediary layer comprises neurons $\mathbf{y}_j$, $j = 1, \ldots, D$ representing D different blur dimensionalities. The output neuron $\mathbf{z}$ initializes the blur estimates by combining the performances of $\mathbf{x}_i$ and $\mathbf{y}_j$. The neurons $\mathbf{x}_i$ and $\mathbf{x}_s$, are instantiated dynamically to $M \times N$ dimensional vectors when connected to $\mathbf{y}_j$ of $M \times N$ blur support.

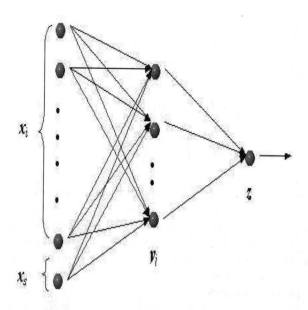

Figure 7.1: Probability neural network for stochastic initialization

The linear summation at $\mathbf{y}_j$ is given by

$$\mathbf{y}_j = \sum_i w_{ji}\mathbf{x}_i + w_{js}\mathbf{x}_s \quad j = 1, \ldots, D \tag{7.7}$$

where w_{ji} is the connection weight between $\mathbf{y}_j$ and $\mathbf{x}_i$, and w_{js} is the connection weight between $\mathbf{y}_j$ and $\mathbf{x}_s$. The output node $\mathbf{z}$ combines $\mathbf{y}_j$, $j = 1, \ldots, D$ to give:

$$\mathbf{z} = \sum_j \delta_{jz}\mathbf{y}_j \tag{7.8}$$

where δ_{jz} is the connection weight between $\mathbf{y}_j$ and $\mathbf{z}$. The network evaluates the performances of $\mathbf{x}_i$, and adjusts the weights w_{ji} and w_{js} to reflect their degree of relevance accordingly. The blur dimensionalities are determined from the nearest neighbor of the 2D mixture Gaussian distribution. The systematic landscape surveying and support estimation provide an efficient approach to achieve intelligent initialization.

Network Weight Estimation

The probability-based weight estimation is formulated in accordance with the notion that subspace with good solutions should be given more emphasis. The weight estimation algorithm for w_{ji} and w_{js} are summarized as follows:

i) Initialize $\mathbf{x}_i, i = 1, \ldots, K$ to blur knowledge base with random parameters, and instantiate them with different dimensionalities of $\mathbf{y}_j$, $j = 1, \ldots, D$ to form the *a priori* subspace, $\mathbf{X}_{ji}$.

ii) For each $\hat{\mathbf{h}}_{ji} \in \mathbf{X}_{ji}$, restore the image $\hat{\mathbf{f}}_{ji}$,

$$\hat{\mathbf{f}}_{ji} = \arg \min_{\hat{\mathbf{f}}} J(\hat{\mathbf{f}}|\mathbf{g}, \hat{\mathbf{h}}_{ji} \in \mathbf{X}_{ji}) \tag{7.9}$$

and evaluate their fidelity measure E_{ji},

$$E_{ji} = \frac{1}{2}\left\|\mathbf{g} - \hat{\mathbf{H}}_{ji}\hat{\mathbf{f}}_{ji}\right\|^2 \tag{7.10}$$

where $\hat{\mathbf{H}}_{ji}$ is the block Toeplitz matrix formed by $\hat{\mathbf{h}}_{ji}$.

iii) Compute the knowledge and stochastic probabilities, P_{ji} and P_{js}:

$$P_{ji} = e^{-\chi\frac{E_{ji}}{\sum_i E_{ji}}} \quad i = 1, \ldots, K \quad j = 1, \ldots, D \tag{7.11}$$

and

$$P_{js} = e^{-\chi\frac{\min\{E_{ji}, \overline{\Delta E}_{ji}\}}{\sum_i E_{ji}}} \quad j = 1, \ldots, D \tag{7.12}$$

where $\overline{\Delta E}_{ji}$ is the average difference between E_{ji} and χ is the scaling coefficient.

iv) Estimate the probability-based connection weights:

$$w_{ji} = \frac{P_{ji}}{\sum\limits_i P_{ji} + P_{js}}, \quad w_{js} = \frac{P_{js}}{\sum\limits_i P_{ji} + P_{js}} \quad i = 1,\ldots,K \quad j = 1,\ldots,D$$

(7.13)

The scheme incorporates the knowledge base, instantiates them with random parameters, evaluates their performances and integrates the feedback into the connection weights. The fidelity error E_{ji} is used as a performance indicator. The probabilities P_{ji} and P_{js} indicate the extent of prior knowledge and stochastic perturbation that should be assigned in the initialization scheme. The weights w_{ji} and w_{js} are the normalized probabilities for P_{ji} and P_{js}, respectively. If the instantiated ϖth blur structure $\mathbf{x}_\varpi$ dominates the others, its fidelity error $E_{j\varpi}$ will be small, leading to a similar range of values for $P_{j\varpi}$ and P_{js}. This combination induces a higher emphasis surrounding the subspace of $\mathbf{x}_\varpi$. On the other hand, if none of the knowledge-based blur instantiations shows good performance, E_{ji} tend to cluster together, leading to a small $\overline{\Delta E}_{ji}$ value. The resulting small P_{ji} and large P_{js} values imply a random initialization where the knowledge base is irrelevant. The asymptotic contribution of the stochastic perturbation ranges from 50% in the first case to 100% in the second case, thereby ensuring the randomness of the initial search space.

The main idea of the new support estimation scheme is centered on modeling the 2D joint probability distribution of the blur supports. We model the blur dimensionality $\mathbf{s} = (s_x, s_y)$ as a 2D mixture Gaussian distribution centered at selected mean $\mathbf{s}_j = (s_{xj}, s_{yj})$, $j = 1, \ldots, D$. This is analogous to function extrapolation in the regression neural network. If a selected dimension $\mathbf{s}_j$ shows good performance, its probability distribution is strengthened, resulting in more emphasis surrounding its support dimensionality. A novel entropy-based performance measure is introduced to assess the likelihood of $\mathbf{s}_j$, and their performances are consequently encoded into the connection weights. The support dimensionality $\mathbf{s}$ is determined by firing the neuron that corresponds to the nearest neighbor. The algorithm of the weight estimation for δ_{jz} is summarized below:

i) Initialize $\mathbf{x}_i$ to random blur and $\mathbf{y}_j$ to different dimensions to form the support subspace,

$$\mathbf{S} = \{\mathbf{s_j}|\mathbf{s_j} = (s_{xj}, s_{yj})\} \quad j = 1,\ldots,D \qquad (7.14)$$

ii) For each $\hat{\mathbf{h}}_j \in \mathbf{S}$, restore the image $\hat{\mathbf{f}}_j$, and evaluate the support estimation measure E_j:

$$\hat{\mathbf{f}}_j = \arg \min_{\hat{\mathbf{f}}} J(\hat{\mathbf{f}}|\mathbf{g}, \hat{\mathbf{h}}_j \in \mathbf{S}) \qquad (7.15)$$

$$E_j = \frac{1}{2}\left\|\mathbf{g} - \hat{\mathbf{H}}_j \hat{\mathbf{f}}_j\right\|^2 + \frac{1}{2}\mathbf{u}^T \mathbf{W}_u \mathbf{u} + \frac{1}{2}\mathbf{v}^T \mathbf{W}_v \mathbf{v} \qquad (7.16)$$

where $\mathbf{u}$ and $\mathbf{v}$ are the entropy and inverse-entropy vector, and $\mathbf{W}_u$ and $\mathbf{W}_v$ are the corresponding weight matrices.

iii) Compute the support probability, P_j:

$$P_j = e^{-\chi \frac{E_j}{\sum_j E_j}} \qquad (7.17)$$

iv) Model the support estimate **s** with a 2D mixture Gaussian distribution centered at $\mathbf{s}_j$:

$$\mathbf{s} = \frac{\sum_j P_j \mathbf{N}(\mathbf{s}_j, \sigma_j^2)}{\sum_j P_j} \qquad (7.18)$$

where $\mathbf{N}(\mathbf{s}_j, \sigma_j^2)$ is a 2D Gaussian distribution with mean $\mathbf{s}_j = (s_{xj}, s_{yj})$ and variance $\sigma_j^2 = (\sigma_{xj}^2, \sigma_{yj}^2)$. The two elements variance σ_j^2 is given by:

$$\sigma_j = \mathbf{d} \frac{E_j}{\sum_j E_j} \qquad (7.19)$$

where **d** is the mean Euclidean separation between $\mathbf{s}_j$

v) Determine the firing neuron z:

$$z = \arg \min_j \|\mathbf{s} - \mathbf{s}_j\| \qquad (7.20)$$

vi) Compute the support weight δ_{jz}:

$$\delta_{jz} = \begin{cases} 1 & \text{if } z = j \\ 0 & \text{if } z \neq j \end{cases} \qquad (7.21)$$

where δ_{jz} is the Kronecker delta function.

The performance measure E_j employs a complementary pair of entropy-based criteria to estimate blur dimensionality. Figure 7.2 provides an illustration of the blur support estimation. The original "Lena" image in Figure 7.2a was degraded by a 5×5 Gaussian mask with a standard deviation of 2.0 to form Figure 7.2b. If the selected dimension $\mathbf{s}_j$ is smaller than the actual support, hardly any textured or edge details are restored as shown in Figure 7.2c, leading to a costly inverse-entropy term. Conversely, if $\mathbf{s}_j$ is larger than the actual support, extreme ringing and noise amplification dominates the scene as in Figure 7.2d, leading to a significant entropy term. Their compromise induces a higher probability in achieving good blur support initialization, exemplified by Figure 7.2e. It is clear that the restored image based on the random mask with good support dimension in Figure 7.2e outperforms those with undersized or oversized estimations in Figures 7.2c and d, respectively. The ringing artifact in Figure 7.2e that arises from random initialization will improve throughout the blind deconvolution process in later stages. The formulation of the performance measure will be explained thoroughly below.

(a)   (b)

(c)   (d)

(e)

Figure 7.2: Illustrations of blur dimensionality estimation. (a) Original Lena image. (b) Image degraded by 5 × 5 Gaussian blur with a standard deviation of 2.0. (c) Restored image using an undersized 3 × 3 random mask. (d) Restored image using an oversized 7 × 7 random mask. (e) Restored image using a 5 × 5 random mask

7.2.2 Knowledge-Based Reinforced Mutation

Dynamic Mutation Attractors

Studies show that most real-life blurs satisfy up to a certain degree of parametric structures. These include motion blur, out-of-focus blur, uniform blur, pillbox uniform blur, Gaussian blur, and sinc-square blurs, among others. Traditional blind methods require a hard decision on whether or not the blurs satisfy a parametric structure prior to problem formulation. This is elusive as the blur structures are often unknown *a priori*. To address this difficulty, we integrate the blur information into a solution space with embedded dynamic mutation attractors. This encourages a soft learning approach where relevant information is incorporated into the blur identification continually. A maximum *a posteriori* (MAP) estimator is devised to determine these attractors:

$$\vec{\mathbf{h}} = \arg\max_{\tilde{\mathbf{h}} \in \tilde{H}} L(\tilde{\mathbf{h}}|\hat{\mathbf{h}}) \tag{7.22}$$

$$= \arg\max_{\tilde{\mathbf{h}} \in \tilde{H}} \log p(\tilde{\mathbf{h}}|\hat{\mathbf{h}}) \tag{7.23}$$

where $\tilde{H}$ is the knowledge-based solution space, $L(\tilde{\mathbf{h}}|\hat{\mathbf{h}})$ is the log likelihood function and $p(\tilde{\mathbf{h}}|\hat{\mathbf{h}})$ is the conditional probability density function of $\tilde{\mathbf{h}}$ given the observation $\hat{\mathbf{h}}$. Assuming that $n = \tilde{\mathbf{h}} - \hat{\mathbf{h}}$ follows a multivariate Gaussian distribution, we can rewrite (7.22) in terms of the covariance matrix, Σ_{nn}:

$$\vec{\mathbf{h}} = \arg\max_{\tilde{\mathbf{h}} \in \tilde{H}} \log p(\tilde{\mathbf{h}}|\hat{\mathbf{h}})$$

$$= \arg\max_{\tilde{\mathbf{h}} \in \tilde{H}} \log \left\{ \frac{1}{(\sqrt{2\pi})^{MN} |\Sigma_{nn}|^{\frac{1}{2}}} e^{-\frac{1}{2}(\tilde{\mathbf{h}}-\hat{\mathbf{h}})^{\mathrm{T}} \Sigma_{nn}^{-1} (\tilde{\mathbf{h}}-\hat{\mathbf{h}})} \right\} \tag{7.24}$$

$$= \arg\max_{\tilde{\mathbf{h}} \in \tilde{H}} \left\{ -\frac{1}{2}MN\log(2\pi) - \frac{1}{2}\log|\Sigma_{nn}| - \frac{1}{2}(\tilde{\mathbf{h}} - \hat{\mathbf{h}})^{\mathrm{T}} \Sigma_{nn}^{-1} (\tilde{\mathbf{h}} - \hat{\mathbf{h}}) \right\} \tag{7.25}$$

If the blur coefficients are uncorrelated, we can further simplify (7.24) by letting $\Sigma_{nn} = \sigma_n^2 \mathbf{I}$:

$$\vec{\mathbf{h}} = \arg\max_{\tilde{\mathbf{h}} \in \tilde{H}} \left\{ -\frac{1}{2}MN\log(2\pi) - \frac{1}{2}\log(\sigma_n^2 MN) - \frac{1}{2\sigma_n^2}(\tilde{\mathbf{h}} - \hat{\mathbf{h}})^{\mathrm{T}} (\tilde{\mathbf{h}} - \hat{\mathbf{h}}) \right\}$$

$$= \arg\min_{\tilde{\mathbf{h}} \in \tilde{H}} \left\{ \frac{1}{2}MN\log(2\pi) + \frac{1}{2}\log(\sigma_n^2 MN) + \frac{1}{2\sigma_n^2}(\tilde{\mathbf{h}} - \hat{\mathbf{h}})^{\mathrm{T}} (\tilde{\mathbf{h}} - \hat{\mathbf{h}}) \right\} \tag{7.26}$$

The MAP estimator determines the structure and support of the mutation attractors from $\tilde{H}$. The noise variance, σ_n^2 has little effect on the estimator especially for small or known blur dimension.

The construction of solution space $\tilde{H}$ involves the incorporation of the blur knowledge. The main feature of the technique includes its ability to integrate any prior experience or devise any targeted structures without compromising the flexibility of the scheme. The solution space $\tilde{\mathrm{H}}$ can be expressed as:

$$\tilde{\mathrm{H}} = \bigcup_i \tilde{\mathrm{H}}_i = \bigcup_i \left\{ \tilde{\mathbf{h}} : \tilde{\mathbf{h}} = \varphi_i(\mathbf{s}_i, \theta_i) \right\} \tag{7.27}$$

where $\tilde{\mathrm{H}}_i$ is the blur subspace, φ_i is the 2D blur function, $\mathbf{s}_i$ is the support, and θ_i is the defining parameters for ith parametric structure. In this chapter, we will focus on 2D blur structures for general purposed image deconvolution. Several well-known blurs such as uniform blur, Gaussian blur and concentric linear blurs are chosen for the stochastic initialization and attractor space construction. The respective subspace of the uniform, Gaussian and linear concentric blurs are given by $\tilde{\mathrm{H}}_u$, $\tilde{\mathrm{H}}_g$ and $\tilde{\mathrm{H}}_l$:

$$\tilde{\mathrm{H}}_u = \left\{ \tilde{h} : \tilde{h}(x,y) = \frac{1}{MN} \quad x = 1,\ldots,M \quad y = 1,\ldots,N \right\} \tag{7.28}$$

$$\tilde{\mathrm{H}}_g = \left\{ \tilde{h} : \tilde{h}(x,y) = Ce^{-\frac{x^2+y^2}{2\sigma^2}} \quad x = 1,\ldots,M \quad y = 1,\ldots,N \right\} \tag{7.29}$$

$$\tilde{\mathrm{H}}_l = \left\{ \tilde{h} : \tilde{h}(x,y) = p\sqrt{x^2+y^2} + q \quad x = 1,\ldots,M \quad y = 1,\ldots,N \right\} \tag{7.30}$$

where (M, N) is the blur support, σ and C are the standard deviation and normalizing constant of the Gaussian blur, and p, q are the gradient and central peak value of the linear blur. Uniform blur is the 2D extension of 1D motion blur, and is characterized completely by its dimension. The Gaussain blur is widely observed in applications such as X-ray imaging, and is difficult to estimate using traditional blur identification approaches. The 2D linear blur is implemented as the first-order estimation to the generic blur.

The construction of the solution space $\tilde{\mathrm{H}}$ also includes constraints to preserve the average intensity of the original image by imposing the unity and non-negativity constraints on $\tilde{\mathrm{H}}$:

$$\sum_x \sum_y h(x,y) = 1 \tag{7.31}$$

$$h(x,y) \geq 0 \tag{7.32}$$

Reinforced Mutation in Attractor Space

Conventional evolutionary strategies employ mutation to provide a biased random walk through the solution space. The common disadvantage of this approach lies in their slow convergence, particularly for high-dimensional problems such as image restoration. To address this difficulty, a novel reinforced mutation scheme that integrates classical mutation and reinforced learning is introduced.

The new technique offers a compromise between stochastic search through traditional mutation, and pattern acquisition through reinforced learning. If the
evolved blur shows a close proximity to the attractor, suggesting a strong likelihood that the actual blur indeed satisfies a high degree of parametric structures,
reinforced learning toward the attractors is emphasized. Otherwise, classical
mutation is employed to explore the search space. The reinforced mutation
paradigm integrates the blur knowledge domain to improve the convergence of
the scheme, leading to significant reduction in the computational cost.

The reinforced mutation operator $\mathbf{M}$ involves the functional mapping of
$\mathbf{M}{:}\Re^{M \times N} \to \Re^{M \times N}$ given by:

$$\hat{\mathbf{h}}_k^{i+1} = \mathbf{M}\left(\hat{\mathbf{h}}_k^i\right) \quad k = 1, \ldots, \mu \tag{7.33}$$

where $\hat{\mathbf{h}}_k^i$ and $\hat{\mathbf{h}}_k^{i+1}$ are the kth blur estimate for ith and $(i+1)$th generations.
The functional mapping can be decoupled into two reinforced mutation equations
described by:

$$\hat{\mathbf{h}}_k^{i+1} = \hat{\mathbf{h}}_k^i + (1-\alpha)\Delta\hat{\mathbf{h}}_k^i - \alpha \, \frac{\partial L(\hat{\mathbf{h}}_k^i, \tilde{\mathbf{h}}(\vec{\theta}_k^i))}{\partial \hat{\mathbf{h}}_k^i} \tag{7.34}$$

$$\vec{\theta}_k^i = \tilde{\theta}_k^i + a^i \Delta \tilde{\theta}_k^i - \beta \, \frac{\partial L(\tilde{\theta}_k^i, \theta_k^{i-1})}{\partial \tilde{\theta}_k^i} \tag{7.35}$$

The first decoupling equation provides the fundamental reinforced learning pro-
cess, where $\Delta\hat{\mathbf{h}}_k^i$ is the stochastic perturbation, $\tilde{\mathbf{h}}(\theta_k^i)$ is the predictor for the
dynamic attractor, α and $L(\hat{\mathbf{h}}_k^i, \tilde{\mathbf{h}}(\theta_k^i))$ are the blur-domain learning rate and
cost function. The second decoupling equation functions as the parametric vec-
tor estimation that characterizes the attractors. θ_k^i is the reinforced parametric
vector, $\tilde{\theta}_k^i$ is the MAP parametric vector, $\Delta\tilde{\theta}_k^i$ is the stochastic perturbation,
$\Delta\theta_k^{i-1}$ is the best vector obtained up to $(i-1)$th generation, β and $L(\tilde{\theta}_k^i, \theta_k^{i-1})$
are the parametric-domain learning rate and cost function.

The new scheme estimates the dynamic attractors and their defining parametric vectors, assesses their relevance with respect to the computed blurs and
provides pattern acquisition through reinforced mutation based on the relevance
of the estimates. If $\hat{\mathbf{h}}_k^i$ evolves within the vicinity of the attractor $\tilde{\mathbf{h}}(\theta_k^i)$, the
likelihood that it actually assumes the parametric structure of $\tilde{\mathbf{h}}(\theta_k^i)$ is significant. In these cases, pattern acquisition toward the predictor is encouraged, and
stochastic search is weakened. The reverse scenario applies if the evolved blurs
are located farther away from the mutation attractors.

The classical mutation $\Delta\hat{\mathbf{h}}_k^i$ in (7.34) can be expressed in terms of its strategic
vectors γ_k^{i-1} by:

$$\Delta\hat{\mathbf{h}}_k^i = \mathbf{N}\left(0, \gamma_k^{i-1}\right) = \left(\gamma_{k,1}^{i-1} e^{z_0+z_1}, \ldots, \gamma_{k,MN}^{i-1} e^{z_0+z_{MN}}\right)^T \tag{7.36}$$

where the variables z_0 and z_n provide constant and variable perturbations to
γ_k^{i-1} and are given by

$$z_0 = \mathbf{N}\left(0, \tau_0^2\right) \tag{7.37}$$

$$z_n = \mathrm{N}\left(0, \tau_n^2\right) \quad n = 1, \ldots, MN \tag{7.38}$$

The typical values of τ_0 and τ_n given by (7.39) and (7.40) are adopted in this work:

$$\tau_0 = \frac{1}{\sqrt{2MN}} \tag{7.39}$$

$$\tau_n = \frac{1}{\sqrt{2\sqrt{2MN}}} \tag{7.40}$$

The corresponding mutation in the parametric domain $\Delta\tilde{\theta}_k^i$ can be expressed in the similar fashion:

$$\Delta\tilde{\theta}_k^i = \mathrm{N}\left(0, \eta_k^{i-1}\right) = \left(\eta_{k,1}^{i-1} e^{z_0+z_1}, \ldots, \eta_{k,R}^{i-1} e^{z_0+z_R}\right)^T \tag{7.41}$$

where η_k^{i-1} is the strategic vector, and R is the dimensionality of $\Delta\tilde{\theta}_k^i$.

The relevance of the attractor $\tilde{\mathbf{h}}(\theta_k^i)$ with respect to the evolved blur $\hat{\mathbf{h}}_k^i$ is captured by the learning rate α:

$$\alpha = \alpha_0 e^{-\xi\left\|\tilde{\mathbf{h}}(\tilde{\theta}_k^i) - \hat{\mathbf{h}}_k^i\right\|^2} \tag{7.42}$$

where ξ is the attractor field strength, and α_0 is the upper acquisition threshold. The field strength coefficient ξ defines the range of vicinity and the extent of learning, while the threshold α_0 is used to ensure that a minimum degree of stochastic exploration is available throughout the process.

The predictive scheme for the attractors in (7.35) incorporates the previous vector estimation into the steepest descent learning term. The stochastic mutation and experienced-based reinforced learning form a complementary pair to explore the search space efficiently. The MAP vector $\tilde{\theta}_k^i$, estimated from the parametric space Θ is given by:

$$\tilde{\theta}_k^i = \arg\max_{\theta \in \Theta} \log p(\tilde{\mathbf{h}}(\theta) | \hat{\mathbf{h}}_k^i) \tag{7.43}$$

The experiences of the previous estimations are captured by the momentum term as:

$$\widehat{\theta}_k^{i-1} = \min_{j=1}^{i-2}\{\mathrm{E}[\tilde{\mathbf{h}}(\tilde{\theta}_k^j)]\} \tag{7.44}$$

where $\mathrm{E}(.)$ is the performance measure given by (7.16). The momentum term is integrated into the learning rate β, which functions as a relevance measure between $\tilde{\theta}_k^i$ and θ_k^{i-1}. The coefficient β, expressed in term of the performance measure and scaling factor ω is given by:

$$\beta = 1 - e^{-\omega\left\|E(\tilde{\theta}_k^i) - E(\widehat{\theta}_k^{i-1})\right\|^2} \tag{7.45}$$

If the performance of θ_k^{i-1} is superior to that of $\tilde{\theta}_k^i$, namely, $E(\widehat{\theta}_k^{i-1})$ is much smaller than $E(\tilde{\theta}_k^i)$, we should emphasize the momentum term to take advantage of the previous good estimations. On the other hand, if their performances are comparable, no obvious advantage is gained through the previous experiences. In these cases, we will reduce the reinforced learning and encourage the classical stochastic search instead.

An important criterion in the formulation of cost function $L(\hat{\mathbf{h}}_k^i, \tilde{\mathbf{h}}(\vec{\theta}_k^i))$ and $L(\tilde{\theta}_k^i, \widehat{\theta_k^{i-1}})$ is their proximity interpretation in the multidimensional vector space. The Euclidean distance measure is adopted due to its simplicity and proximity-discerning capability. The cost functions $L(\hat{\mathbf{h}}_k^i, \tilde{\mathbf{h}}(\tilde{\theta}_k^i))$ and $L(\tilde{\theta}_k^i, \theta_k^{i-1})$ are given by:

$$L(\hat{\mathbf{h}}_k^i, \tilde{\mathbf{h}}(\tilde{\theta}_k^i)) = \frac{1}{2}\left\|\tilde{\mathbf{h}}(\tilde{\theta}_k^i) - \hat{\mathbf{h}}_k^i\right\|^2 \tag{7.46}$$

$$L(\tilde{\theta}_k^i, \widehat{\theta_k^{i-1}}) = \frac{1}{2}\left\|\tilde{\theta}_k^i - \widehat{\theta_k^{i-1}}\right\|^2 \tag{7.47}$$

An interesting feature of the new approach is its ability to provide parametric blur abstraction. This can be achieved by providing a generic parametric model such as a second-order quadratic function, or targeted structure if there is any prior experience, to the blur knowledge domain. The idea of parametric abstraction is particularly important if we would like to generalize a class of imaging processes whose blurring characteristics are unknown. The new approach will render an abstraction if the actual blur satisfies up to a certain degree of parametric structure, subject to our modeling. Even if the actual blur differs from any of the parametric structures, the obtained abstraction may function as an alternative blur model to facilitate problem formulations and simplifications.

The new technique makes a trade-off between the convergence efficiency and the chances that the actual blur may lie near to, but not at the parametric structure. Fortunately, this dilemma can be alleviated in two ways. First, the upper learning threshold α_0 ensures that a minimum degree of stochastic mutation is always available, thereby providing a certain degree of search space surrounding the attractor. Second, even if the parametric blur is obtained as an abstraction for the actual blur, their close proximity will guarantee a good image restoration. In addition, the computational cost of achieving such a satisfactory deconvolution is reduced considerably.

7.2.3 Perception-Based Image Restoration

The image-domain solution space Ω_i given in (7.5) is generated based on the evolved blur population Φ_i:

$$\Omega_i = \left\{\hat{\mathbf{f}}_i | \hat{\mathbf{f}}_i = \arg\min_{\hat{\mathbf{f}}} J(\hat{\mathbf{f}}|\mathbf{g}, \hat{\mathbf{h}}_i \in \Phi_i)\right\} \tag{7.48}$$

The cost function $J(\hat{\mathbf{f}}|\mathbf{g}, \hat{\mathbf{h}}_i \in \Phi_i)$ is the convex set projection of $J(\hat{\mathbf{f}}, \hat{\mathbf{h}}|\mathbf{g})$ onto the image subspace, and is given by:

$$J(\hat{\mathbf{f}}|\mathbf{g}, \hat{\mathbf{h}}_i \in \Phi_i) = \frac{1}{2}\|\mathbf{g} - \hat{\mathbf{H}}_i\hat{\mathbf{f}}\|^2 + \frac{1}{2}\hat{\mathbf{f}}^T\Lambda\mathbf{D}^T\mathbf{D}\hat{\mathbf{f}} + C_f \tag{7.49}$$

where Λ is the image-domain regularization matrix, $\mathbf{D}$ is the Laplacian high-pass operator, and C_f is the projection constant. The cost function consists of a data-conformance and a model-conformance term. The data-conformance term measures the fidelity error of the restored image. However, the term alone is sensitive to singularities, and tends to form undesirable ringing and noise amplification across regions with abrupt visual activities. Therefore, the model-conformance term or regularization functional is introduced to provide a smoothing constraint on the restored image.

A main issue in the formulation of the cost function involves providing a fine balance between the fidelity term and regularization functional. We adopt an adaptive, perception-based regularization matrix Λ to control their relative contributions:

$$\Lambda = \mathrm{diag}(\lambda_1, \lambda_2, \ldots, \lambda_{PQ}) \tag{7.50}$$

where $P \times Q$ is the image dimension, and λ_i is the regularization coefficient for the ith image pixel, and is given as:

$$\lambda_i = \max\left\{\lambda_l, -0.5[\lambda_h - \lambda_l]\log \sigma_i^2 + \lambda_h\right\} \tag{7.51}$$

The coefficients λ_l and λ_h are the lower and upper regularization values, and σ_i^2 is the local variance of ith pixel. λ_i is assigned as a decreasing logarithmic function of σ_i^2 because the textured and edge image pixels that exhibit high local variance require a small regularization parameter to preserve their fine details. The reverse holds true for the smooth regions that prefer noise and ringing suppression. The logarithmic function is adopted as it is used commonly to model the non-linear transformation of human visual perception.

Constrained conjugate-gradient optimization is chosen to minimize (7.49) due to its lower computational complexity and storage requirement as compared to Quasi-Newton, and faster convergence with more robustness with respect to steepest-descent technique. As the underlying Toeplitz distortion matrix is highly sparse, the convergence of the solution can be achieved very quickly. In practical applications, the algorithm is terminated when convergence or a predetermined number of iterations is reached. The $(P \times Q)$-dimensional gradient of $J(\hat{\mathbf{f}}|\mathbf{g}, \hat{\mathbf{h}}_i \in \Phi_i)$ is denoted as:

$$\mathbf{J}_{\hat{\mathbf{f}}} = \begin{bmatrix} J_{\hat{\mathbf{f}}_1} & \cdots & J_{\hat{\mathbf{f}}_{PQ}} \end{bmatrix}^{\mathrm{T}} \tag{7.52}$$

where $J_{\hat{\mathbf{f}}_i}$ is the partial derivative of $J(\hat{\mathbf{f}}|\mathbf{g}, \hat{\mathbf{h}}_i \in \Phi_i)$ with respect to $\hat{f}_i$ and is given by:

$$
\begin{aligned}
J_{\hat{\mathbf{f}}_i} &= \frac{\partial J(\hat{\mathbf{f}}|\mathbf{g}, \hat{\mathbf{h}}_i \in \Phi_i)}{\partial \hat{f}_i} \\
&= -\left\{(g_i - h_i * f_i) * h_{-i}\right\} + \lambda_i\left\{(d_i * f_i) * d_{-i}\right\}
\end{aligned} \tag{7.53}
$$

where g_i, h_i, f_i, d_i and λ_i are the corresponding scalar entries of $\mathbf{g}$, $\mathbf{h}$, $\mathbf{f}$, $\mathbf{D}$ and Λ, respectively.

The mathematical formulations of image restoration based on conjugate gradient optimization are summarized as follows. To simplify the notation, the indices of the ith blur estimates are assumed implicitly.

i) Initialize the conjugate vector, $\mathbf{q}$ with the gradient, $(\mathbf{J}_{\hat{\mathbf{f}}})_0$:

$$\mathbf{q}_0 = (\mathbf{J}_{\hat{\mathbf{f}}})_0 \tag{7.54}$$

ii) Calculate the kth iteration's adaptive step size to update the image:

$$\eta_k = \frac{\|\mathbf{q}_k\|_\Gamma^2}{\|\mathbf{H}_k \mathbf{q}_k\|_\Gamma^2 + \mathbf{q}_k^T \Lambda \mathbf{D}^T \mathbf{D} \mathbf{q}_k} \tag{7.55}$$

where $\mathbf{H}_k$ is the Toeplitz matrix formed by $\mathbf{h}_k$ and Γ is the image-domain support.

iii) Compute the updated image:

$$\hat{\mathbf{f}}_{k+1} = \hat{\mathbf{f}}_k + \eta_k \mathbf{q}_k \tag{7.56}$$

iv) Calculate the kth iteration's adaptive step size to update the conjugate vector:

$$\rho_k = \frac{\left\| (\mathbf{J}_{\hat{\mathbf{f}}})_{k+1} \right\|_\Gamma^2}{\left\| (\mathbf{J}_{\hat{\mathbf{f}}})_k \right\|_\Gamma^2} \tag{7.57}$$

v) Update the new conjugate vector:

$$\mathbf{q}_{k+1} = -\left(\mathbf{J}_{\hat{\mathbf{f}}}\right)_{k+1} + \rho_k \mathbf{q}_k \tag{7.58}$$

vi) Impose image constraint to preserve the consistency of the display range:

$$0 \le \hat{f}_{k+1}(i,j) \le L \quad \forall (i,j) \in \Gamma \tag{7.59}$$

where $L = 2^b$ is the highest gray level, and b is the number of bits used in image representation.

vii) Repeat steps (ii) to (vi) until convergence or a maximum number of iterations is reached

7.2.4 Recombination Based on Niche-Space Residency

Many classical evolutionary schemes employ recombination to generate offspring by combining two or more randomly chosen candidates [40, 156]. These techniques include global and local intermediary recombination, uniform crossovers and γ-point crossovers. They often emphasize the randomness of recombination, with little consideration for the behaviors of the solutions. In this section, we introduce a new recombination technique based on subspace residency of the

candidates. The underlying principle is that the solutions sharing a uniquely defined subspace should be reinforced, while those from different subspaces should
be encouraged to recombine freely.

We introduce a new unique subspace called *niche-space* N_k, that describes the
spatial association of the solutions. The niche-space of each solution is centered
at $\mathbf{n}_\nu$ with the identifier ν given by:

$$\nu = \min_s \left\| \hat{\mathbf{h}} - \arg \max_{\tilde{\mathbf{h}} \in \tilde{\mathbf{H}}_s} \log p(\tilde{\mathbf{h}}|\hat{\mathbf{h}}) \right\| \tag{7.60}$$

$$\mathbf{n}_\nu = \arg \max_{\tilde{\mathbf{h}} \in \tilde{\mathbf{H}}_\nu} \log p(\tilde{\mathbf{h}}|\hat{\mathbf{h}}) \tag{7.61}$$

The identifier ν represents the *a priori* blur type with $\mathbf{n}_\nu$ being the MAP estimate
from its class. The niche-space is defined by:

$$N_v = \left\{ \hat{\mathbf{h}} : d(\hat{\mathbf{h}}, \mathbf{n}_\nu) \le \varepsilon \right\} \tag{7.62}$$

where ε is the niche-space radii, and $d(\hat{\mathbf{h}}, \mathbf{n}_\nu)$ expressed in term of the scaling
coefficient χ, is given by:

$$d(\hat{\mathbf{h}}, \mathbf{n}_\nu) = e^{-\chi \left\| \hat{\mathbf{h}} - \mathbf{n}_\nu \right\|^2} \tag{7.63}$$

The niche-space is a region of interest because it describes the subspace where
the parametric blurs are likely to be found. Therefore, solutions originating
from the same niche-space should be strengthened to enhance their existence.
In contrast, classical recombination is employed to recombine solutions coming
from different niche-spaces. The recombination operator $\mathbf{R}:\Re^{M \times N} \rightarrow \Re^{M \times N}$
involves the selection of random candidates, their niche-space determination,
and recombination according to their residency.

The new recombination algorithm is summarized as follows:

For $n = 1$ to μ

begin

- Generate two random candidates: $\mathbf{h}_i$, $\mathbf{h}_j$, $i, j \quad \in 1, \ldots, \mu$

- If their support size differs, namely $\mathbf{s}_i \ne \mathbf{s}_j$, choose either of the candidates:

    $\mathbf{h}_r = \text{random}\,(\mathbf{h}_i,\, \mathbf{h}_j)$

    Else if $\mathbf{h}_i,\, \mathbf{h}_j \in \mathbf{N}_\nu$, perform intra-niche-space recombination:

    $\mathbf{h}_r = R_{intra}\,(\mathbf{h}_i,\, \mathbf{h}_j)$

    Else, perform inter-niche-space recombination:

$$\mathbf{h}_r = R_{inter}\left(\mathbf{h}_i,\ \mathbf{h}_j\right)$$

end

The inter-niche-space recombination operator, R_{inter} employs global intermediary recombination to give:

$$\mathbf{h_r} = \frac{1}{2}\left(\mathbf{h}_i + \mathbf{h}_j\right) \tag{7.64}$$

On the other hand, the intra-niche-space recombination operator, R_{intra} is computed using:

$$\mathbf{h}_r = R_{intra}\left(\mathbf{h}_i, \mathbf{h}_j\right) = \frac{d\left(\mathbf{h}_i, \mathbf{n}_{\nu i}\right)\mathbf{h}_i + d\left(\mathbf{h}_j, \mathbf{n}_{\nu j}\right)\mathbf{h}_j}{d\left(\mathbf{h}_i, \mathbf{n}_{\nu i}\right) + d\left(\mathbf{h}_j, \mathbf{n}_{\nu j}\right)} \tag{7.65}$$

If $\hat{\mathbf{h}}_i$ is close the to $\mathbf{n}_{\nu i}$, $d(\hat{\mathbf{h}}_i, \mathbf{n}_{\nu i})$ will have a large contribution, leading to a stronger $\hat{\mathbf{h}}_i$ emphasis in the weighted blur estimate. The reverse applies if $\hat{\mathbf{h}}_i$ and $\mathbf{n}_{\nu i}$ are farther apart. The new scheme is developed based on the principle that the evolved blurs that originate from a similar niche-space, namely, sharing similar *a priori* blur structure, should be enhanced. In comparison, solutions that reside in different niche-spaces are encouraged to recombine freely as in classical recombination.

7.2.5 Performance Evaluation and Selection

Algorithmic tractability has always been the core consideration in most optimization problems. This is because conventional methods usually require these preconditions to enable their formulations. For instance, the gradient-based optimization methods such as steepest descent and Quasi-Newton techniques require the cost function to be differentiable. They fail to function properly if the preconditions are not satisfied. In addition, the chosen cost function may not be the best criterion that provides the most tangible interpretation to the problem. In these circumstances, evolutionary optimizations become an instrumental tool as they allow the most tangible performance measure to be adopted.

In view of this, an entropy-based objective function $F{:}\Re^{P\times Q + M\times N} \rightarrow \Re$ is introduced that functions as the restoration performance indicator:

$$F(\hat{\mathbf{h}}, \hat{\mathbf{f}}) = \frac{1}{2}\left\|\mathbf{g} - \hat{\mathbf{H}}\hat{\mathbf{f}}\right\|^2 + \frac{1}{2}\mathbf{u}^T\mathbf{W}_u\mathbf{u} + \frac{1}{2}\mathbf{v}^T\mathbf{W}_v\mathbf{v} \tag{7.66}$$

where $\mathbf{u}$ and $\mathbf{v}$ are the entropy and inverse-entropy vectors, and $\mathbf{W}_u$ and $\mathbf{W}_v$ are the corresponding weight matrices. The new objective function utilizes the intuitive local entropy statistics to assess the performance of the restored images. The entropy and inverse-entropy vectors are the lexicographically ordered visual activity measure for the local image neighborhood. They can be represented by:

$$\mathbf{u} = \left[\ u_1\ \ \cdots\ \ u_{PQ}\ \right]^T \tag{7.67}$$

$$\mathbf{v} = \left[\ v_1\ \ \cdots\ \ v_{PQ}\ \right]^T \tag{7.68}$$

where u_i and v_i are the complementary entropy and inverse-entropy pair defined by:

$$u_i = -\sum_j p_j \log p_j \tag{7.69}$$

$$v_i = \log n - \sum_j p_j \log p_j \tag{7.70}$$

The intensity probability histogram p_j is used to evaluate u_i and v_i, with n representing the number of classes in the histogram. The entropy u_i has larger values in high visual activity areas such as textured and edge regions. On the other hand, the inverse-entropy v_i has larger values in the low visual activity regions such as smooth backgrounds. The weight matrices $\mathbf{W}_u$ and $\mathbf{W}_v$ are given by:

$$\mathbf{W_u} = \operatorname{diag}(w_{u1}, w_{u2}, \ldots, w_{uPQ}) \tag{7.71}$$
$$\mathbf{W_v} = \operatorname{diag}(w_{v1}, w_{v2}, \ldots, w_{vPQ}) \tag{7.72}$$

The weighted coefficients w_{ui} and w_{vi} present an adaptive cost-penalty effect, and is described by:

$$w_{ui} = \alpha\left\{1 - \tanh\left(\frac{\sigma_i^2 - m}{m}\right)\right\} \tag{7.73}$$

$$w_{vi} = \beta\left\{1 + \tanh\left(\frac{\sigma_i^2 - m}{m}\right)\right\} \tag{7.74}$$

where α and β are the upper entropy and inverse-entropy emphasis thresholds, and m is the transition threshold between the smooth and textured regions. The shifted hyperbolic tangential function is adopted due to its negative symmetric property centered at the transition threshold m. The weighted matrices are used in conjunction with entropy and inverse-entropy vectors to ensure detail recovery in the textured regions, and ringing suppression in the smooth backgrounds.

The weight matrix $\mathbf{W}_u$ has large values in the smooth backgrounds and small values at the textured regions. This is combined with an entropy vector to ensure that ringing and noise amplification is penalized in smooth backgrounds. The reverse argument applies for the weight matrix $\mathbf{W}_v$. It has small weights at smooth backgrounds and large weights at textured regions. Together with the inverse-entropy vector, it encourages detail recovery in the textured regions by penalizing low visual activities in the respective areas. The incorporation of the superior entropy-based performance measure is made feasible owing to the virtue of evolutionary strategies, as its nondifferentiablity will restrict its application in traditional optimization methods.

A combination of deterministic and stochastic selection $\mathbf{S}{:}\Re \rightarrow \Re^{P \times Q + M \times N}$ is adopted to choose the offspring for the future generation. In accordance with classical evolutionary strategies, deterministic selection is employed to propagate those candidates with superior performance into the future generations. To complement the deterministic process, stochastic selection such as tournament selection is adopted to improve the diversity and randomness of the solutions. The combination of deterministic and stochastic selections ensures the convergence of the algorithm by the virtue of convergence theorem given in [160].

The evolutionary process continues until convergence or the maximum number of iterations is reached. The termination criterion is satisfied when the relative reduction in the objective function falls below a predefined threshold ε:

$$|F(T) - F(T-1)| < \varepsilon |F(T)| \qquad (7.75)$$

where $F(T)$ and $F(T-1)$ are the objective functions for the Tth and $(T-1)$th generations, respectively.

7.3 Soft-Decision Method

7.3.1 Recursive Subspace Optimization

As with the CRL method, we incorporate regularization in the soft-decision method. The new cost function that comprises a data fidelity measure, image and blur-domain regularization terms, and the soft-decision blur estimation error is given as

$$J(\hat{\mathbf{h}}, \hat{\mathbf{f}}) = \tfrac{1}{2}||\mathbf{g} - \hat{\mathbf{H}}\hat{\mathbf{f}}||^2 + \tfrac{1}{2}\hat{\mathbf{f}}^{\mathrm{T}}\Lambda\mathbf{D}^{\mathrm{T}}\mathbf{D}\hat{\mathbf{f}} + \tfrac{1}{2}\hat{\mathbf{h}}^{\mathrm{T}}\Psi\mathbf{E}^{\mathrm{T}}\mathbf{E}\hat{\mathbf{h}} + \tfrac{1}{2}\tilde{p}||\mathbf{w}^{\mathrm{T}}(\hat{\mathbf{h}}-\tilde{\mathbf{h}})||^2 \quad (7.76)$$

where Λ and Ψ are the image and blur-domain regularization matrices, $\mathbf{D}$ and $\mathbf{E}$ are the Laplacian high-pass operators, $\tilde{p}$ is the soft-decision proximity measure, $\mathbf{w}$ is the weighted blur emphasis, and $\tilde{\mathbf{h}}$ is the soft parametric blur estimate. The data fidelity term functions as a restoration criterion. However, the measure alone is sensitive to abrupt changes in visual activities, and tends to form undesirable ringing over smooth image backgrounds. Therefore, the image and blur-domain regularization terms are incorporated into the cost function to lend stability into the system. The image-domain regularization term encapsulates the regularization matrix Λ, to provide an adaptive image restoration. Likewise, the blur-domain regularization term includes the regularization matrix Ψ, to render piece-wise smoothness in the blur. The blur estimation error determines the proximity measure and modeling error of $\hat{\mathbf{h}}$ with respect to the soft estimate, $\tilde{\mathbf{h}}$ generated by the estimator. If there exists a high degree of proximity, the confidence that the current blur assumes a parametric structure increases, leading to a gradual adaptation of $\hat{\mathbf{h}}$ toward the soft estimate. Otherwise, the fidelity and regularization terms will dominate the restoration criteria and the soft estimation error will wither away.

Due to the distinctive characteristics and properties of $\hat{\mathbf{h}}$ and $\hat{\mathbf{f}}$, the optimization objectives and priorities for them should be handled differently. Therefore, it is both algorithmically and intuitively sensible to project $J(\hat{\mathbf{h}}, \hat{\mathbf{f}})$ into the image and blur subspaces to form the following cost functions:

$$J(\hat{\mathbf{f}}|\hat{\mathbf{h}}) = \frac{1}{2}||\mathbf{g} - \hat{\mathbf{H}}\hat{\mathbf{f}}||^2 + \frac{1}{2}\hat{\mathbf{f}}^{\mathrm{T}}\Lambda\mathbf{D}^{\mathrm{T}}\mathbf{D}\hat{\mathbf{f}} + C_f \qquad (7.77)$$

$$J(\hat{\mathbf{h}}|\hat{\mathbf{f}}) = \tfrac{1}{2}\|\mathbf{g} - \hat{\mathbf{H}}\hat{\mathbf{f}}\|^2 + \tfrac{1}{2}\hat{\mathbf{h}}^{\mathbf{T}}\mathbf{\Psi}\mathbf{E}^{\mathbf{T}}\mathbf{E}\hat{\mathbf{h}} + \tfrac{1}{2}\tilde{p}\|\mathbf{w}^{\mathbf{T}}(\hat{\mathbf{h}}-\tilde{\mathbf{h}})\|^2 + C_h \qquad (7.78)$$

where C_f and C_h are the projection constants of J $(\hat{\mathbf{h}}, \hat{\mathbf{f}})$ onto the image and blur domains, respectively.

The restored image, $\hat{\mathbf{f}}$, and identified blur, $\hat{\mathbf{h}}$, are estimated by minimizing the cost functions in their respective subspaces. An alternating minimization (AM) process is employed to optimize $J(\hat{\mathbf{f}}|\hat{\mathbf{h}})$ and $J(\hat{\mathbf{h}}|\hat{\mathbf{f}})$ recursively, thereby improving the $\hat{\mathbf{f}}$ and $\hat{\mathbf{h}}$ estimates progressively. The algorithmic simplicity of AM enables the scheme to project and minimize the image and blur cost functions one at a time.

The mathematical formulations of recursive subspace optmization are summarized as follows, with Γ and H denoting the solution spaces of $\hat{\mathbf{f}}$ and $\hat{\mathbf{h}}$, respectively:

i) Initialize $\hat{\mathbf{h}}_0$ to a random mask

ii) For $(i + 1)$th recursion, solve for $\hat{\mathbf{f}}_{i+1}$ and $\hat{\mathbf{h}}_{i+1}$:

$$\hat{\mathbf{f}}_{i+1} = \arg \min_{\hat{\mathbf{f}}\in\Gamma} J(\hat{\mathbf{f}}|\hat{\mathbf{h}}_i) \qquad (7.79)$$

$$\hat{\mathbf{h}}_{i+1} = \arg \min_{\hat{\mathbf{h}}\in\mathrm{H}} J(\hat{\mathbf{h}}|\hat{\mathbf{f}}_{i+1}) \qquad (7.80)$$

iii) Stop when the convergence or the maximum number of iterations is reached

This algorithm differs from the symmetrical double regularization (SDR) approaches suggested by You and Kaveh [153], and Chan and Wong [154] in several ways. Their investigations assume that blind image deconvolution can be decomposed into two symmetrical processes of estimating the image and blur. This assumption conflicts with several observations. First, most PSFs exist in the form of low-pass filters. In contrast, typical images consist of smooth, texture and edge regions, whose frequency spectrums vary considerably from low in the smooth regions to mediocre and high in the texture and edge regions. Second, image restoration is inherently a perception-based process as the quality of the restored image relies heavily upon the visual inspection of humans. Third, most real-life PSFs satisfy up to a certain degree of parametric structures. In comparison, we have little or no prior knowledge about most imaging scenes. These observations illustrate the importance of performing image restoration and blur identification in accordance to their priorities and characteristics. The new approach attempts to address these asymmetries by integrating parametric blur information into the scheme, and tailors image restoration and blur identification in accordance to their unique properties.

7.3.2   Hierarchical Neural Network for Image Restoration

Structure and Properties of Hierarchical Cluster Model

The formation of each degraded image pixel, g_{ij}, involves the filtering operation of a f_{ij} neighborhood window by a spatially finite PSF. As a result, only a finite local neighborhood window of g_{ij} is required to estimate each f_{ij}. This implies that image restoration is conceptually a collection of local parallel processes with a global coordination. Hierarchical Cluster Model (HCM) is a nested neural network consisting of parallel, distributed sub-networks or clusters. Its distributed nature underscores the unique properties of local pixel formation while its hierarchy provides an overall coordination. HCM can be constructed in a bottom-up manner. Individual neurons form the trivial level 0 clusters of the network. The neurons sharing similar functionality and characteristics coalesce into numerous level 1 clusters. Likewise, those level 1 clusters with homogeneous characteristics in turn coalesce into level 2 clusters. This process continues repeatedly to form a hierarchical neural network with multidimensional distributed clusters.

The sparse intercluster synaptic connections of HCM are effective in reducing the computational cost of restoration. As its cluster structures closely match the homogenous image regions, HCM is uniquely placed to provide a perception-based restoration. The notion of clustering lends tolerance to the image stationarity constraint, as it divides a potentially inhomogeneous image into several quasi-stationary clusters. In addition, the modular structure encourages the developments of visually pleasing homogeneous regions.

Optimization of Image-Domain Cost Function as HCM Energy Minimization

The formulation of $J(\hat{\mathbf{f}}|\hat{\mathbf{h}})$ optimization, as the HCM energy minimization based on a novel regularization matrix Λ, is developed in this section. The image-domain cost function, $J(\hat{\mathbf{f}}|\hat{\mathbf{h}})$ in (7.77) can be expressed as:

$$J(\hat{\mathbf{f}}|\hat{\mathbf{h}}) = \tfrac{1}{2}\hat{\mathbf{f}}^{\mathrm{T}}(\hat{\mathbf{H}}^{\mathrm{T}}\hat{\mathbf{H}} + \Lambda\mathbf{D}^{\mathrm{T}}\mathbf{D})\hat{\mathbf{f}} - (\mathbf{g}^{\mathrm{T}}\hat{\mathbf{H}})\hat{\mathbf{f}} + \tfrac{1}{2}\mathbf{g}^{\mathrm{T}}\mathbf{g} + C_f \qquad (7.81)$$

We can simplify (7.81) by assigning $\mathbf{P}=\hat{\mathbf{H}}^{\mathrm{T}}\hat{\mathbf{H}}$, $\mathbf{Q}=\mathbf{D}^{\mathrm{T}}\mathbf{D}$, $\mathbf{r}=\hat{\mathbf{H}}^{\mathrm{T}}\mathbf{g}$ and ignoring the constants $\tfrac{1}{2}\mathbf{g}^{\mathrm{T}}\mathbf{g}$ and C_f:

$$J(\hat{\mathbf{f}}|\hat{\mathbf{h}}) = \tfrac{1}{2}\hat{\mathbf{f}}^{\mathrm{T}}(\mathbf{P} + \Lambda\mathbf{Q})\hat{\mathbf{f}} - \mathbf{r}^{\mathrm{T}}\hat{\mathbf{f}} \qquad (7.82)$$

The cost function can be decomposed into smooth S, texture T and edge E partitions by introducing:

$$\hat{\mathbf{f}} = \begin{pmatrix} \hat{\mathbf{f}}_s \\ \hat{\mathbf{f}}_t \\ \hat{\mathbf{f}}_e \end{pmatrix},$$

$$\Lambda = \begin{bmatrix} \Lambda_s & \Phi & \Phi \\ \Phi & \Lambda_t & \Phi \\ \Phi & \Phi & \Lambda_e \end{bmatrix},$$

$$\mathbf{P} = \begin{bmatrix} \mathbf{P}_{ss} & \mathbf{P}_{st} & \mathbf{P}_{se} \\ \mathbf{P}_{ts} & \mathbf{P}_{tt} & \mathbf{P}_{te} \\ \mathbf{P}_{es} & \mathbf{P}_{et} & \mathbf{P}_{ee} \end{bmatrix},$$

$$\mathbf{Q} = \begin{bmatrix} \mathbf{Q}_{ss} & \mathbf{Q}_{st} & \mathbf{Q}_{se} \\ \mathbf{Q}_{ts} & \mathbf{Q}_{tt} & \mathbf{Q}_{te} \\ \mathbf{Q}_{es} & \mathbf{Q}_{et} & \mathbf{Q}_{ee} \end{bmatrix},$$

$$\mathbf{r} = \begin{pmatrix} \mathbf{r}_s \\ \mathbf{r}_t \\ \mathbf{r}_e \end{pmatrix} \tag{7.83}$$

where $\hat{\mathbf{f}}_s$, $\hat{\mathbf{f}}_t$, $\hat{\mathbf{f}}_e$ are the lexicographically ordered pixels, and Λ_s, Λ_t, Λ_e are the regularization submatrices for smooth, texture and edge cluster types. The matrices $\mathbf{P}$, $\mathbf{Q}$, and vector $\mathbf{r}$ are partitioned likewise into their corresponding submatrices and subvector. It is observed from (7.83) that there exists a cyclic symmetry between the smooth, texture and edge partitions. Substitute (7.83) into (7.82) and simplify the results using the implicit cyclic symmetry, we obtain:

$$J(\hat{\mathbf{f}}|\hat{\mathbf{h}}) = \sum_\alpha \frac{1}{2}\hat{\mathbf{f}}_\alpha^{\mathrm{T}}(\mathbf{P}_{\alpha\alpha} + \Lambda_\alpha\mathbf{Q}_{\alpha\alpha})\hat{\mathbf{f}}_\alpha + \sum_{\alpha,\beta:\,\alpha\neq\beta}\sum \frac{1}{2}\hat{\mathbf{f}}_\alpha^{\mathrm{T}}(\mathbf{P}_{\alpha\beta} + \Lambda_\alpha\mathbf{Q}_{\alpha\beta})\hat{\mathbf{f}}_\beta - \sum_\alpha \mathbf{r}_\alpha^{\mathrm{T}}\hat{\mathbf{f}}_\alpha$$

$$\tag{7.84}$$

where $\alpha, \beta \in \{S, T, E\}$ are the cluster types. The first and second summation terms in (7.84) represent the contributions from similar and different cluster types, respectively.

Consider the decomposition of (7.84) into the cluster level:

$$\hat{\mathbf{f}}_\alpha = \begin{pmatrix} \hat{\mathbf{f}}_1^\alpha \\ \vdots \\ \hat{\mathbf{f}}_\varphi^\alpha \end{pmatrix},$$

$$\Lambda_\alpha = \begin{bmatrix} \lambda_1^\alpha \mathbf{I} & & \Phi \\ & \ddots & \\ \Phi & & \lambda_\varphi^\alpha \mathbf{I} \end{bmatrix},$$

$$\mathbf{P}_{\alpha\beta} = \begin{bmatrix} \mathbf{P}_{11}^{\alpha\beta} & \cdots & \mathbf{P}_{1\gamma}^{\alpha\beta} \\ \vdots & & \vdots \\ \mathbf{P}_{\varphi 1}^{\alpha\beta} & \cdots & \mathbf{P}_{\varphi\gamma}^{\alpha\beta} \end{bmatrix},$$

$$\mathbf{Q}_{\alpha\beta} = \begin{bmatrix} \mathbf{Q}_{11}^{\alpha\beta} & \cdots & \mathbf{Q}_{1\gamma}^{\alpha\beta} \\ \vdots & & \vdots \\ \mathbf{Q}_{\varphi 1}^{\alpha\beta} & \cdots & \mathbf{Q}_{\varphi\gamma}^{\alpha\beta} \end{bmatrix},$$

$$r_\alpha = \begin{pmatrix} r_1^\alpha \\ \vdots \\ r_\varphi^\alpha \end{pmatrix} \tag{7.85}$$

where φ and γ are the number of clusters for partition types, α and β.

Substitute (7.85) into (7.84) to express $J(\hat{\mathbf{f}}|\hat{\mathbf{h}})$ in terms of cluster levels, we obtain:

$$
J(\hat{\mathbf{f}}|\hat{\mathbf{h}})
$$
$$
= \left\{ \begin{array}{l} \sum_k \sum_\alpha \frac{1}{2}\hat{\mathbf{f}}_k^{\alpha\mathrm{T}}(\mathbf{P}_{kk}^{\alpha\alpha} + \lambda_k^\alpha \mathbf{Q}_{kk}^{\alpha\alpha})\hat{\mathbf{f}}_k^\alpha + \sum_{k,l:\,k\neq l}\sum_\alpha \frac{1}{2}\hat{\mathbf{f}}_k^{\alpha\mathrm{T}}(\mathbf{P}_{kl}^{\alpha\alpha} + \lambda_k^\alpha \mathbf{Q}_{kl}^{\alpha\alpha})\hat{\mathbf{f}}_l^\alpha + \\ \sum_{k,l:\,k\neq l}\sum_{\alpha,\beta:\,\alpha\neq\beta} \frac{1}{2}\hat{\mathbf{f}}_k^{\alpha\mathrm{T}}(\mathbf{P}_{kl}^{\alpha\beta} + \lambda_k^\alpha \mathbf{Q}_{kl}^{\alpha\beta})\hat{\mathbf{f}}_l^\beta + \sum_k \sum_\alpha r_k^{\alpha\mathrm{T}}\hat{\mathbf{f}}_k^\alpha \end{array} \right\} \tag{7.86}
$$

where k, l are the cluster designations. The first term in (7.86) refers to the intracluster contribution. The second and third term correspond to intercluster contributions arising from similar and different cluster types.

The energy function of a three-level HCM is given by

$$
E = -\frac{1}{2}\left\{ \sum_{k=1}^{K}\sum_{i=1}^{Pk}\sum_{j=1}^{Pk} w_{ik,jk}s_{ik}s_{jk} + \sum_{k\neq l}\gamma_{kl}\sum_{i=1}^{Pk}\sum_{j=1}^{Pl} w_{ik,jl}s_{ik}s_{jl} \right\} - \sum_{k=1}^{K}\sum_{i=1}^{Pk} b_{ik}s_{ik} \tag{7.87}
$$

where $s_{ik}, s_{jl} \in \{0,\ldots,L-1\}$ are the state of ith neuron in cluster k and jth neuron in cluster l, $w_{ik,jk}$ is the intracluster connection weight between neurons s_{ik} and s_{jk}, $w_{ik,jl}$ is the intercluster connection weight between neurons s_{ik} and s_{jl}, γ_{kl} is the average strength of intercluster connections to intracluster connections, b_{ik} is the bias input to neuron s_{ik}, $L = 2^b$ is the number of gray levels, and b is the number of bits used in pixel representation. There are K first-level partitions with P_k and P_l representing the number of neurons in cluster k and cluster l, respectively. The first and second summation triplets in $\{.\}$ represent the intracluster and intercluster contributions to the overall energy function.

A close inspection reveals the correspondence between the image-domain cost function in (7.86) and the energy function of HCM in (7.87). The HCM energy function is expressed in terms of generic clusters from smooth, texture and edge partitions. Comparing these two expressions by mapping neuron s to pixel $\hat{f}$, and extending the generic clusters in (7.87) to encapsulate the detailed clusters in (7.86), we obtain:

$$w_{ik,jk}^{\alpha\alpha} = -\left(p_{kk}^{\alpha\alpha}\right)_{ij} - \lambda_k^\alpha\left(q_{kk}^{\alpha\alpha}\right)_{ij} \qquad \alpha\in\{S,T,E\} \tag{7.88}$$

$$w_{ik,jl}^{\alpha\alpha} = -\left(p_{kl}^{\alpha\alpha}\right)_{ij} - \lambda_k^\alpha\left(q_{kl}^{\alpha\alpha}\right)_{ij} \qquad \alpha\in\{S,T,E\} \tag{7.89}$$

$$w_{ik,jl}^{\alpha\beta} = -\left(p_{kl}^{\alpha\beta}\right)_{ij} - \lambda_k^\alpha\left(q_{kl}^{\alpha\beta}\right)_{ij} \qquad \alpha,\beta\in\{S,T,E\} \tag{7.90}$$

$$b_{ik} = r_{ik} \tag{7.91}$$

$$\gamma_{kl} = 1 \tag{7.92}$$

where $w_{ik,jk}^{\alpha\alpha}$, $w_{ik,jl}^{\alpha\alpha}$ and $w_{ik,jl}^{\alpha\beta}$ are the intracluster connection weight of cluster type α, intercluster connection weight within cluster type α, and intercluster connection weight between cluster types α and β. $(p_{kk}^{\alpha\alpha})_{ij}$, $(q_{kk}^{\alpha\alpha})_{ij}$, $(p_{kl}^{\alpha\alpha})_{ij}$, $(q_{kl}^{\alpha\alpha})_{ij}$, $\left(p_{kl}^{\alpha\beta}\right)_{ij}$, $\left(q_{kl}^{\alpha\beta}\right)_{ij}$ are the ij-th scalar entries of submatrices $\mathbf{P}_{kk}^{\alpha\alpha}$, $\mathbf{Q}_{kk}^{\alpha\alpha}$, $\mathbf{P}_{kl}^{\alpha\alpha}$, $\mathbf{Q}_{kl}^{\alpha\alpha}$, $\mathbf{P}_{kl}^{\alpha\beta}$, $\mathbf{Q}_{kl}^{\alpha\beta}$, respectively. We can express (7.88) to (7.91) in terms of image degradation parameters by restoring $\mathbf{P} = \hat{\mathbf{H}}^\mathsf{T}\hat{\mathbf{H}}$, $\mathbf{Q} = \mathbf{D}^\mathsf{T}\mathbf{D}$, and $\mathbf{r} = \hat{\mathbf{H}}^\mathsf{T}\mathbf{g}$ to give

$$w_{ik,jk}^{\alpha\alpha} = -\left[\sum_{n=1}^{N}\hat{h}_{ni}\hat{h}_{nj}\right]_{kk}^{\alpha\alpha} - \lambda_k^\alpha\left[\sum_{n=1}^{N}d_{ni}d_{nj}\right]_{kk}^{\alpha\alpha} \tag{7.93}$$

$$w_{ik,jl}^{\alpha\alpha} = -\left[\sum_{n=1}^{N}\hat{h}_{ni}\hat{h}_{nj}\right]_{kl}^{\alpha\alpha} - \lambda_k^\alpha\left[\sum_{n=1}^{N}d_{ni}d_{nj}\right]_{kl}^{\alpha\alpha} \tag{7.94}$$

$$w_{ik,jl}^{\alpha\beta} = -\left[\sum_{n=1}^{N}\hat{h}_{ni}\hat{h}_{nj}\right]_{kl}^{\alpha\beta} - \lambda_k^\alpha\left[\sum_{n=1}^{N}d_{ni}d_{nj}\right]_{kl}^{\alpha\beta} \tag{7.95}$$

$$b_{ik} = \left[\sum_{n=1}^{N}\hat{h}_{ni}g_n\right]_k \tag{7.96}$$

where N is the number of neurons in the network. The parameter γ_{kl} is unity as the space-invariance of PSF causes the intercluster synaptic strength to be as strong as the intracluster synaptic strength. The regularization parameter of kth cluster for tth AM recursion, $\lambda_k(t)$ is given by:

$$\lambda_k(t) = \max\left\{\lambda_l(t), -0.5[\lambda_h(t) - \lambda_l(t)]\log \sigma_k^2 + \lambda_h(t)\right\} \qquad (7.97)$$

where σ_k^2 is the average local variance of kth cluster, and $\lambda_l(t)$, $\lambda_h(t)$ are the lower and upper regularization thresholds given by:

$$\lambda_l(t) = [\lambda_l(0) - \lambda_l(\infty)]e^{-t} + \lambda_l(\infty) \qquad (7.98)$$

$$\lambda_h(t) = [\lambda_h(0) - \lambda_h(\infty)]e^{-t} + \lambda_h(\infty) \qquad (7.99)$$

$\lambda_l(0)$, $\lambda_h(0)$, $\lambda_l(\infty)$, $\lambda_h(\infty)$ are the lower and upper regularization values in the beginning and final AM recursion. $\lambda_k(t)$ is assigned as a decreasing logarithm function of σ_k^2 because the texture and edge clusters that have high local variance require a small regularization parameter to preserve the fine details. The reverse holds true for the smooth clusters. The logarithm function is adopted as it commonly used to model the non-linear transformation of human visual perception. In general, $\lambda_l(0)$ and $\lambda_h(0)$ are comparatively greater than $\lambda_l(\infty)$ and $\lambda_h(\infty)$. This enables the restoration scheme to recover more details progressively as we become more confident about the identified blur and the restored image.

Cluster Formation and Restoration

A three-level HCM is constructed from the blurred image in two stages. In the first stage, the image is segmented into smooth, texture and edge regions. These regions are further partitioned into separate homogeneous clusters in the second stage. The mathematical representations for the cluster formation are:

$$I = S \cup T \cup E \qquad (7.100)$$

$$S = \bigcup_i^x S_i \quad T = \bigcup_j^y T_j \quad E = \bigcup_k^z E_k \qquad (7.101)$$

where S_i, T_j and E_k are the ith smooth cluster, jth texture cluster and kth edge cluster, respectively.

The blurred image is segmented into different regions based on its local statistics. The standard Sobel operator is used to extract the edges as they exhibit a distinctive change of gray levels compared to blurred texture regions. To further differentiate smooth backgrounds from texture regions, the blurred image is uniformly restored for an iteration using a random mask with estimated dimension of Section 7.2.1. We observe that smooth backgrounds in the partially restored image are mostly monotonous with low visual activities. Therefore, several visual measures such as contrast, entropy, and local variance are extracted from the image. An expert voting technique is employed to discriminate whether a pixel belongs to smooth or texture regions. These regions are further divided into non-neighboring clusters to preserve their homogeneity. Throughout HCM formation, small fragmented clusters that arise due to misclassification are continually incorporated into larger neighboring clusters by morphological operations. Our investigation shows that multivalue neuron modeling with steepest gradient descent is both robust and efficient in restoring the image. The neuron updating rule in 2D image indices is given by:

$$\Delta s(x, y) = - \frac{1}{w_{(x,y)}^{(x,y)}} \sum_{(x,y) \in \Re} w_{(x+m,y+n)}^{(x,y)} s(x + m, y + n) + b(x,y) \qquad (7.102)$$

$x = 1, \ldots, P \quad y = 1, \ldots, Q$

where $\Delta s(x, y)$ is the update increment, $w_{(x+m,y+n)}^{(x,y)}$ is the connection weight from neuron $s(x+m, y + n)$ to $s(x, y)$, $b(x, y)$ is the self-bias, and P, Q are the image dimensions. Equation (7.32) reveals that the update of a neuron or pixel value involves a weight kernel $\Re$, of dimension $(2M - 1) \times (2N - 1)$ for an $M \times N$ blur size. This implies a transformation from the initial high dimensional optimization problem to a smaller system with a time complexity of $O(MNPQ)$ per iteration, thereby reducing the computational cost of restoration. At the end of each iteration, the cluster boundaries are adjusted dynamically according to the mean and average local variance of each cluster. This renders a continual improvement in the segmentation map, its cluster structure and, most importantly, the quality of the restored image.

7.3.3   Soft Parametric Blur Estimator

We will use the same soft parametric estimator to model the computed blur with the best-fit parametric structure from a predefined solution space $\tilde{H}$. Therefore, the MAP estimator given in equations (7.22 to 7.26) applies. Again the solution subspaces is the union of uniform, $\tilde{H}_u$, Gaussian, $\tilde{H}_g$ and concentric linear blur, $\tilde{H}_l$ given in equations (7.28 to 7.30), respectively,

$$\tilde{H} = \bigcup_i \tilde{H}_i = \tilde{H}_u \cup \tilde{H}_g \cup \tilde{H}_l \qquad (7.103)$$

As the parametric blurs can be uniquely determined by their defining parameters, we derive a gradient-based approach to estimate these parameters, thereby further improving the processing speed. The simplified parametric estimators are given as:

$$\sigma^2 = \mathrm{E} \left(\frac{\hat{h}r}{|\nabla \hat{h}|} \right)_{\mathrm{H}} \qquad (7.104)$$

where $\mathrm{E}(\, . \,)_{\mathrm{H}}$ is the expectation operator spanning over H, r is the radial distance of the $\hat{h}$ coefficient from the mask center, and $|\nabla \hat{h}|$ is the magnitude of gradient at $\hat{h}$.

The gradient p and central peak value q of the linear blur are estimated by:

$$p = \mathrm{E}\left(|\nabla \hat{h}|\right)_{\mathrm{H}} \tag{7.105}$$

Imposing the unity constraint (7.31) on the linear blur, we obtain the q estimate:

$$q = \frac{E\left(|\nabla \hat{h}|\right)}{MN} \sum_x \sum_y \sqrt{x^2 + y^2} \tag{7.106}$$

An important issue of the estimator is to evaluate the relevance of the parametric estimate to the computed blur. Therefore, a proximity measure is devised as the confidence criteria in our $\tilde{\mathbf{h}}$ estimation. If the current $\hat{\mathbf{h}}$ resembles one of the parametric structures in the solution space, a close proximity will be achieved that enhances our confidence in the $\tilde{\mathbf{h}}$ estimation. The proximity measure is given as

$$\tilde{p} = \begin{cases} 0 & \text{if } \sum_i \left(\tilde{\mathbf{h}}_i - \hat{\mathbf{h}}_i\right)^2 > T \\ A e^{-\xi(\tilde{\mathbf{h}} - \hat{\mathbf{h}})^{\mathrm{T}}(\tilde{\mathbf{h}} - \hat{\mathbf{h}})} & \text{if } \sum_i \left(\tilde{\mathbf{h}}_i - \hat{\mathbf{h}}_i\right)^2 \leq T \end{cases} \tag{7.107}$$

where A and ξ are the scaling factors, and T is the matching threshold.

The proximity measure corresponds to two scenarios. The first scenario describes those irregular $\hat{\mathbf{h}}$ that do not resemble any parametric structures. As their chances of assuming a parametric structure are very slim, a zero value is appropriately assigned to these cases. On the other hand, if a satisfactory match occurs between $\hat{\mathbf{h}}$ and $\tilde{\mathbf{h}}$, a likelihood-based proximity measure is employed. We used a threshold T of 0.02 and ξ of 500 in our experiments. The scale factor, A, is determined by ensuring the contribution of soft estimation term is an order less than the fidelity or regularization term for medium matches, and the same order for close matches. Experimental simulations show that it is unnecessary to evaluate the optimal A value as long as it has the right order of magnitude.

7.3.4 Blur Identification by Conjugate Gradient Optimization

The mathematical objective of blur identification can be obtained by combining (7.78) and (7.80) to give:

$$\hat{\mathbf{h}}_{i+1} = \arg \min_{\hat{\mathbf{h}} \in H} \left\{ \tfrac{1}{2}||\mathbf{g} - \hat{\mathbf{H}}\hat{\mathbf{f}}||^2 + \tfrac{1}{2}\hat{\mathbf{h}}^{\mathrm{T}}\boldsymbol{\Psi}\mathbf{E}^{\mathrm{T}}\mathbf{E}\hat{\mathbf{h}} + \tfrac{1}{2}\tilde{p}||\mathbf{w}^{\mathrm{T}}(\hat{\mathbf{h}} - \tilde{\mathbf{h}})||^2 \right\} \tag{7.108}$$

The cost function consists of three criteria, namely, the data conformance measure, the blur-domain regularization term and the soft-decision blur estimation error. The data conformance and regularization terms ensure the fidelity and piecewise smoothness of the solutions. Their relative contribution is controlled by the regularization matrix, $\boldsymbol{\Psi} = \mathrm{diag}(\psi_1, \psi_2, \ldots, \psi_{MN})$ with

$$\psi_i = [\psi_h - \psi_l]e^{-\sigma_i^2} + \psi_l \quad \forall i \in \mathrm{H} \tag{7.109}$$

where σ_i^2 is the local variance at $\hat{h}_i$, ψ_l and ψ_h are the lower and upper limit of the blur-domain regularization parameter.

The parametric blur information is incorporated into the cost function as the soft estimation error. Its functionality is analogous to Hebbian learning in neural networks where $\tilde{p}$ is the adaptive learning step size, and $||\mathbf{w}^T(\hat{\mathbf{h}}-\tilde{\mathbf{h}})||^2$ is the weighted learning error. The proximity measure, $\tilde{p}$, renders a compromise between the estimation error and the rest of the criteria in $J\left(\hat{\mathbf{h}}|\hat{\mathbf{f}}\right)$. If the correlation between $\hat{\mathbf{h}}$ and the estimate $\tilde{\mathbf{h}}$ improves significantly, $\tilde{p}$ will become dominant and gradually locks $\hat{\mathbf{h}}$ into one of the parametric structures. Otherwise, the soft error will wither away and the combination of fidelity and regularization terms will determine the $\hat{\mathbf{h}}$ estimation. The flexibility of the approach lies in its ability to incorporate and adjust the relevance of the parametric structure throughout the restoration.

The raised cosine function, $\boldsymbol{w}$, of the soft error assigns weighted emphasis to different regions of the blur. The inner coefficients of $\tilde{\mathbf{h}}$ and $\hat{\mathbf{h}}$ are emphasized because estimates of circumferential coefficients are prone to distortions by undesirable ringing in the restored image, particularly in the early stage of restoration. The raised cosine function given in terms of the roll-off factor, α, is:

$$w(x,y) = \begin{cases} \frac{1}{2}\left\{1 + \cos\left[\frac{\pi}{\alpha}\left(\sqrt{\frac{x^2}{M^2} + \frac{y^2}{N^2}} - \frac{1-\alpha}{2}\right)\right]\right\} & \sqrt{\frac{x^2}{M^2} + \frac{y^2}{N^2}} \geq \frac{1-\alpha}{2} \\ 1 & \text{otherwise} \end{cases}$$

(7.110)

Conjugate gradient optimization is again chosen to minimize (7.112) for similar reasons in minimizing the image cost function in CRL. The convergence of an $(M \times N)$-dimensional blur can be achieved in a maximum of $M \times N$ steps. However, faster convergence is expected when the eigenvalues of the Hessian matrix are clustered. In practical applications, the algorithm is terminated when the convergence or the maximum number of iterations is reached. We extend the conjugate gradient approach in [153] to incorporate the soft estimation error in our blur identification. The $(M \times N)$-dimensional gradient of $J\left(\hat{\mathbf{h}}|\hat{\mathbf{f}}\right)$ is denoted as:

$$\mathbf{J_{\hat{h}}} = \begin{bmatrix} J_{\hat{h}_1} \\ \vdots \\ J_{\hat{h}_{MN}} \end{bmatrix}$$

(7.111)

where $J_{\hat{h}_i}$ is the partial derivative of $J\left(\hat{\mathbf{h}}|\hat{\mathbf{f}}\right)$ with respect to $\hat{h}_i$, $i \in \mathrm{H}$ and is given by:

$$J_{\hat{h}_i} = \frac{\partial J(\hat{\mathbf{h}}|\hat{\mathbf{f}})}{\partial \hat{h}_i}$$

$$= -\left\{(g_i - h_i * f_i) * f_{-i}\right\} + \mu_i\{(e_i * h_i) * e_{-i}\} + \tilde{p}w_i\left(\tilde{h}_i - \hat{h}_i\right)$$ (7.112)

where g_i, h_i, f_i, e_i, μ_i and w_i are the corresponding scalar entries of $\mathbf{g}$, $\mathbf{h}$, $\mathbf{f}$, $\mathbf{E}$, Ψ and $\mathbf{w}$, respectively.

The mathematical formulations of blur identification based on conjugate gradient optimization are summarized as follows:

i) Initialize the conjugate vector, $\mathbf{q}$, with the gradient, $(\mathbf{J}_{\hat{\mathbf{h}}})_0$:

$$\mathbf{q}_0 = (\mathbf{J}_{\hat{\mathbf{h}}})_0 \tag{7.113}$$

ii) Calculate the kth iteration's adaptive step size to update the blur:

$$\eta_k = \frac{\|\mathbf{q}_k\|_{\mathrm{H}}^2}{\left\|\mathbf{Q}_k\hat{\mathbf{f}}\right\|_{\Gamma}^2 + \mathbf{q}_k^T \mathbf{\Psi} \mathbf{E}^T \mathbf{E}\mathbf{q}_k + \tilde{p}\mathbf{q}_k^T \mathbf{W}\mathbf{q}_k} \tag{7.114}$$

where $\mathbf{Q}_k$ is the Toeplitz matrix formed by $\mathbf{q}_k$ and $\mathbf{W} = \mathrm{diag}\,(w_1, w_2, \ldots, w_{MN})$

iii) Compute the updated blur:

$$\hat{\mathbf{h}}_{k+1} = \hat{\mathbf{h}}_k + \eta_k\mathbf{q}_k \tag{7.115}$$

iv) Calculate the kth iteration's adaptive step size to update the conjugate vector:

$$\rho_k = \frac{\left\|(\mathbf{J}_{\hat{\mathbf{h}}})_{k+1}\right\|_{\mathrm{H}}^2}{\left\|(\mathbf{J}_{\hat{\mathbf{h}}})_{k}\right\|_{\mathrm{H}}^2} \tag{7.116}$$

v) Update the new conjugate vector:

$$\mathbf{q}_{k+1} = -\left(\mathbf{J}_{\hat{\mathbf{h}}}\right)_{k+1} + \rho_k\mathbf{q}_k \tag{7.117}$$

vi) Repeat steps (ii) to (v) until convergence or a maximum number of iterations is reached

To ensure the realizability of the blurs in most physical applications, we impose the unity and non-negativity constraints in (7.31) and (7.32) on the blur estimate. In addition, the blur is assumed to be centrosymmetric:

$$h(x, y) = h(-x, -y) \quad \forall (x,y) \in \mathrm{H} \tag{7.118}$$

These assumptions confine the solution space to those blurs that we are interested in. For example, the unity constraint in (7.31) is a consequence of similar overall intensity levels between the degraded and original images. The centrosymmetric condition is usually satisfied by most physical PSFs. In general, these constraints improve the robustness and convergence of the scheme.

7.3.5   Blur Compensation

Typical images are comprised of smooth, texture and edge regions. The monotonous smooth regions are ineffectual in recovering the blur. The task of blur estimation relies mainly on the information available in the texture and edge regions. In this section, we will investigate the effect of texture and edge orientation on blur estimation, and devise a compensation scheme to account for the necessary adjustment.

The scalar equivalence of (7.1), ignoring the additive noise, is given by:

$$g_i = f_i * h_i = \sum_{j \in \mathbf{H}} f_{i-j} h_j \tag{7.119}$$

where H is the 2D blur support. If a segment of f assumes an edge structure with F_b and F_e corresponding to the background and edge gray levels, respectively, we can rewrite (7.119) as

$$G_e = F_e \sum_{j \in \mathbf{E}} h_j + F_b \sum_{j \in \mathbf{B}} h_j \tag{7.120}$$

where $\mathbf{E}$ and $\mathbf{B}$ are the sets of blur coefficients overlying on top of the edge and backgrounds, and G_e is the gray level of the corresponding edge in $\mathbf{g}$. Imposing the unity constraint on (7.120) and further simplifying it, we obtain:

$$\sum_{j \in \mathbf{E}} h_j = \frac{G_e - F_b}{F_e - F_b} \tag{7.121}$$

Equation (7.121) reveals the ambiguity of the solutions as the blur cannot be uniquely determined. As the blur coefficients along the edge are indistinguishable, their estimations from these edges will result in statistical averaging. For an image degraded by non-uniform blur that contains a dominant edge orientation, the averaging effect of blur coefficients is stronger along the dominant axis compared to the minor axis. The implication on non-uniform blur, such as the Gaussian mask, is that the central peak coefficients will be smeared out, the major axis coefficients boosted, and the minor axis coefficients weakened. It is worth noting that the extent of these effects depend upon the degree of dominance between the perpendicular major and minor axis. The more asymmetric these two axes are, the more prominent is the effect. In addition, these effects do not affect the uniform blur.

In view of the potential distortion to the estimated blur, a correlation-based approach is developed to compensate for the blur. As correlation is a good measure of edges and textures, its incorporation enables a systematic approach to the blur compensation. We introduce a new neighborhood correlation matrix, Z, that characterizes the correlation of neighborhood image pixels throughout the whole image:

$$Z = \begin{bmatrix} \zeta_{\left(-\frac{M-1}{2},-\frac{N-1}{2}\right)} & \cdots & \zeta_{\left(-\frac{M-1}{2},\frac{N-1}{2}\right)} \\ \vdots & \zeta_{(0,0)} & \vdots \\ \zeta_{\left(\frac{M-1}{2},-\frac{N-1}{2}\right)} & \cdots & \zeta_{\left(\frac{M-1}{2},\frac{N-1}{2}\right)} \end{bmatrix} \tag{7.122}$$

where $\zeta_{(i,j)}$ is defined as

$$\zeta_{(i,j)} = \frac{\sum\limits_{k} \mathbf{R_{ff}}(k,k+i+jM)}{\frac{1}{MN}\sum\limits_{i}\sum\limits_{j}\sum\limits_{k} \mathbf{R_{ff}}(k,k+i+jM)} \tag{7.123}$$

with $\mathbf{R}_{ff}(i,j)$ being the (i,j)th entry of the correlation matrix, $\mathbf{R_{ff}} = \mathrm{E}\left[\hat{\mathbf{f}}\hat{\mathbf{f}}^{\mathrm{T}}\right]$.

The neighborhood correlation matrix provides useful information on the overall correlation structure of the image spanning over the blur support dimension. In particular, the higher-order correlation of $\mathbf{Z}$ along the vertical and horizontal directions indicates the extent of edges or hairy textures in the respective axes. Therefore, the information can be extracted and incorporated into the compensation scheme.

We present a modified 2D Gaussian function, $c(x,y)$, to perform the compensation as below:

$$\hat{h}_c(x,y) = c(x,y) \times \hat{h}(x,y) \quad \forall (x,y) \in \mathrm{H} \tag{7.124}$$

where $\hat{h}_c(x,y)$ is the compensated blur, and $c(x,y)$ is given by:

$$c(x,y) = Be^{-\gamma S(\nabla E_h)\left(\frac{x^2}{2\sigma_x^2}+\frac{y^2}{2\sigma_y^2}\right)} \quad \forall (x,y) \in \mathrm{H} \tag{7.125}$$

The factor γ functions as the compensation coefficient, ∇E_h is the gradient of blur radial energy profile, σ_x^2 and σ_y^2 are the horizontal and vertical variances, B is the normalizing constant, and S(.) is the squash function. A modified Gaussian function is adopted because it is effective in rectifying the distortions, namely, strengthening the central peak coefficient, weakening the major axis coefficients and boosting the minor axis coefficients. The emphasis or de-emphasis of the coefficients are controlled by the vertical and horizontal variances, σ_x^2 and σ_y^2:

$$\sigma_x^2 = \frac{\sum\limits_{i} \sigma_i^2}{\frac{1}{M}\sum\limits_{i}\left(\prod\limits_{j} \zeta_{ij}\right)} \tag{7.126}$$

$$\sigma_y^2 = \frac{\sum\limits_{j} \sigma_j^2}{\frac{1}{N}\sum\limits_{j}\left(\prod\limits_{i} \zeta_{ij}\right)} \tag{7.127}$$

where σ_i^2 and σ_j^2 are the intravariance of ith row and jth column of matrix $\mathbf{Z}$. Both the numerator and denominator in (7.130) and (7.131) are characterizations of edges. If the image is dominated by vertical edges, the intravariance along the column is smaller and the product of neighborhood coefficients along the column is larger. The combined effects cause σ_y^2 to be smaller than σ_x^2, therefore de-emphasizing the coefficients along the dominant vertical axis and emphasizing the coefficients along the horizontal axis. The reverse holds true when the image is dominated by horizontal edges.

The squash function, $S(\nabla E_h)$, takes the radial energy profile of the PSF as:

$$S(\nabla E_h) = \tanh\left(E[\|\nabla E_h\|]\right) = E[\tanh\left(\|\nabla E_h\|\right)] \tag{7.128}$$

A hyperbolic tangent function is employed because its value ranges from zero for uniform blur, up to one for other blurs with decreasing radial energy profiles. This corresponds to negligible compensation for the uniform blur, and desirable correction for the others. The normalizing constant, B, is introduced to ensure the unity condition in (7.31) is satisfied. The compensation factor, γ, can be evaluated by adjusting the extent of correction using:

$$\frac{c(0,0)}{c\left(\frac{M-1}{2}, \frac{N-1}{2}\right)} \leq \beta \tag{7.129}$$

where β is the upper limit for the ratio of central peak to corner coefficient. It is observed experimentally that for most images, a compensation of less than 10% is adequate, which translates to a β value of 1.3 approximately. Finally, the support dimension of the blur estimate is pruned if the relative energy of the circumferential coefficients falls below a threshold, namely,

$$E_c < \ell E_T \tag{7.130}$$

where E_c is the energy of the circumferential coefficients on two sides of the blur, E_T is the total energy of all the coefficients, and ℓ is the pruning threshold.

7.4   Simulation Examples

We demonstrate the performance of the two methods by simulation examples, and compare them with the SDR method. In the evolutionary strategy, the following parameters are used in the simulation: $\mu = 10$, $\kappa = \infty$, $\nu = 10$, and $\rho = 2$. Stochastic initialization was performed using the probability neural network. The blur population was recombined based on niche-space residency. Reinforced mutation with $\alpha_0 = 0.8$, $\xi = 150$, $a = 0.8$ and $\omega = 150$ was adopted.

Image restoration based on conjugate gradient optimization with $\lambda_h = 10^{-2}$ and $\lambda_l = 10^{-4}$ was employed to generate the image-domain solution space. The performances of the restored images and blurring functions were evaluated using the entropy-based objective function with $\alpha = \beta = 0.01$. The combination of the deterministic and stochastic selection was used to propagate those solutions with good characteristics into the future generations. The process of blur recombination, reinforced mutation, image restoration, performance evaluation, and evolution selection was allowed to continue until convergence or a predetermined number of generations was reached.

In the soft-decision method, AM was used to restore the image and identify the blur recursively. The blur estimate was initialized to a random mask whose support size was estimated similar to CRL. The pruning procedure with an ℓ value of 0.15 gradually trimmed the blur support to the actual dimension when necessary. A three-level HCM was constructed using (7.97 to 7.103) with $\lambda_l(0) = 0.005$, $\lambda_h(0) = 0.05$, $\lambda_l(\infty) = 0.0001$, $\lambda_h(\infty) = 0.01$, and the cluster restoration was performed to achieve perception-based restoration. On the other hand, conjugate gradient optimization was used to identify the blur by minimizing the blur-domain cost function. The blur-domain regularization parameters are taken as $\psi_l = 10^6$ and $\psi_h = 1.2 \times 10^6$. The AM continued until the convergence or the maximum number of recursion was reached.

The experimental results given in the following subsections cover the blind image deconvolution of various degraded images under different noise levels and blur structures. To provide an objective performance measure, the signal-to-noise ratio improvement is given as

$$\Delta\text{SNR} = 10 \log_{10} \frac{\|\mathbf{g} - \mathbf{f}\|^2}{\left\|\hat{\mathbf{f}} - \mathbf{f}\right\|^2} \tag{7.131}$$

where $\mathbf{g}$, $\mathbf{f}$ and $\hat{\mathbf{f}}$ represent the degraded, original and restored image, respectively. Nevertheless, the best criterion to assess the performance of the restoration schemes remains the human inspection of the restored images. This is consistent with the view that human vision is, fundamentally, a subjective perception.

7.4.1   Identification of 2D Gaussian Blur

The blind deconvolution of images degraded by Gaussian blur is presented in this subsection. Traditional blind methods are ineffectual in identifying Gaussian blur as it exhibits distinctive gradient changes throughout the mask, conflicting with the usual precondition of piecewise smoothness. In addition, it does not exhibit prominent frequency null, resulting in the failure of frequency-based blind deconvolutions. The original "Lena" image in Figure 7.2a has a dimension of 256 × 256 with 256 gray levels. It was degraded by a 5 × 5 Gaussian mask with a standard deviation of 2.0, coupled with some quantization noise to form Figure 7.3a. Applying the two algorithms on the degraded image, we obtained the restored images given in Figures 7.3b and c, respectively. It is observed that the

(a)   (b)

(c)   (d)

Figure 7.3: Blind restoration of image degraded by Gaussian blur with quantization noise. (a) Image degraded by 5×5 Gaussian blur with quantization noise. (b) Restored image using CRL. (c) Restored image using RSD. (d) Restored image using SDR

two methods achieve almost perfect restoration by recovering the visual clarity and sharpness of the image. It suppresses the ringing and noise amplification in the smooth backgrounds, while preserving the fine details in textured and edge regions.

The restored image using the SDR approach proposed in [153] is given in Figure 7.3d. Comparing the results in Figures 7.3b and c, and that in Figure 7.3d, we observe that the SDR restored image does not recover fine details such as the edges and textured feather regions as well as the methods described in this chapter. Furthermore, ringing starts to develop in the smooth backgrounds near the shoulder and the hat regions. The visual observation is confirmed by the ΔSNR improvement as CRL and RSD methods offer an improvement of 6.1 and 3.9 dB, respectively, compared to the 3.5 dB obtained by the SDR approach. The significant ΔSNR improvement by the CRL method illustrates the advantages of incorporating dynamic attractor into the reinforced mutation scheme.

The corresponding blur estimates by CRL and RSD are given in Figures 7.4a

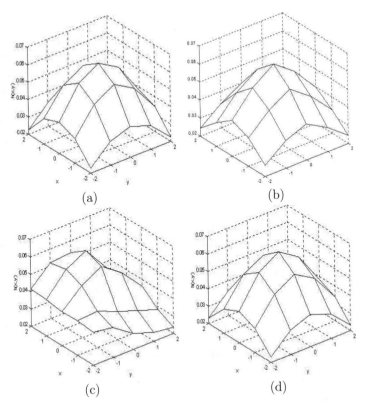

Figure 7.4: Blur estimates for blind deconvolution of image degraded by Gaussian blur. (a) Blur estimate by CRL. (b) Blur estimate by RSD. (c) Blur estimate by SDR. (d) Actual 5×5 Gaussian blur with a standard deviation of 2.0.

and b, respectively. We observe that the blur estimates by the two methods closely resemble the actual Gaussian blur shown in Figure 7.4d. The identified blur obtained using the SDR approach is given in Figure 7.4c. Comparing Figures 7.4a and b with Figure 7.4c, it is clear that SDR fails to identify the blur adequately. In particular, the coefficients in the upper half of the blur receive exceeding weights, while the central peak coefficient is mislocated. The SDR approach emphasizes on the piecewise smoothness in both the image and blur domains, leading to insufficient texture recoveries in the restored image and consequently poor blur identification.

7.4.2 Identification of 2D Gaussian Blur from Degraded Image with Additive Noise

In this subsection, we illustrate the effectiveness of the CRL and RSD to deconvolve degraded image with additive noise. The original image in Figure 7.2a was degraded by the exact 5×5 Gaussian mask, coupled with 30 dB additive noise to form Figure 7.5a. We applied the algorithms on the degraded image to obtain the restored images in Figures 7.5b and c. It is observed that the restored images feature fine details at the textured feather regions while suppressing ringing and noise amplification in the smooth backgrounds. This reflects the fact that the random additive noise has been amplified during the restoration process. The restored image using the SDR approach is given in Figure 7.5d. The comparison between Figures 7.5b and c with Figure 7.5d shows the superior restoration results obtained using the two methods. The SDR restored image does not preserve the fine details near the textured regions. Again, ringing starts to appear particularly in the shoulder and the nearby square background. The observations are consistent with the objective measure as our approaches offer ΔSNR improvements of 3.4 and 2.24 dB, respectively, compared to the 1.28 dB obtained by the SDR method. Comparing the restored image using our approach with and without additive noise in Figures 7.5b and c and Figures 7.3c and d, it is clear that they show comparable image quality. On the other hand, the SDR restored images show noticeable deterioration in the quality of the restored images.

The corresponding blur estimates by CRL and RSD are presented in Figures 7.6a and b, respectively. We can see clearly see that the blur estimates match the actual blur in Figure 7.4d closely. Compared with the identified blur obtained using the SDR approach in Figure 7.6c, it is obvious that CRL and RSD offer superior blur identifications, and consequently render better image restorations.

7.4.3   Identification of 2D Uniform Blur by CRL

To illustrate the flexibility of the new approach in dealing with different blur structures, we deconvolved an image degraded by uniform blur with some additive noise using the evolutionary strategy method. The original Lena image in Figure 7.2a was degraded by a 5×5 uniform blur coupled with 40 dB additive noise to form Figure 7.7a. The restored image after applying the CRL algorithm is given in Figure 7.7b. It is observed that the restored image features detailed textured and edge regions, with no visible ringing in the smooth backgrounds. Compared to the restored image using the SDR approach in Figure 7.7c, it is obvious that our technique achieves better restoration results by rendering details and suppressing ringing in the appropriate image regions. The ΔSNR improvement supports the visual inspection as our approach offers an improvement of 6.4 dB compared to the 5.2 dB obtained by the SDR method.

The evolution of the corresponding blur estimates is given in Figure 7.8. The initial blur estimate, the evolved solution after a few generations, and the final estimate are given in Figures 7.8a, b and c, respectively. The random estimate in Figure 7.8a evolves gradually to the final estimate in Figure 7.8(c), assimilating the good properties as it progresses. We notice that the final blur estimate matches the actual uniform blur in Figure 7.8e closely. The identified blur using the SDR approach in Figure 7.8d achieves a reasonably good estimation, albeit at the cost of strong smoothing constraints in the blur and image domains. This leads to insufficient detail recovery in the textured and edge regions shown in Figure 7.7c. In addition, the excessive smoothing constraint prohibits successful

(a)   (b)

(c)   (d)

Figure 7.5: Blind deconvolution of image degraded by Gaussian blur with additive noise. (a) Image degraded by 5×5 Gaussian blur with 30 dB additive noise. (b) Restored image using CRL. (c) Restored image using RSD. (d) Restored image using SDR

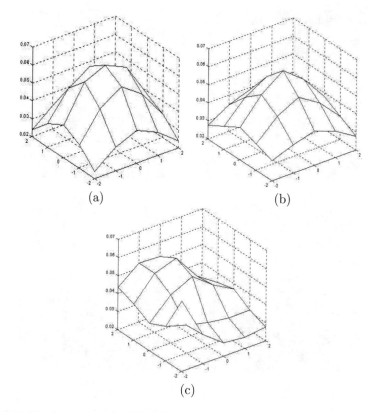

Figure 7.6: Blur estimates for blind deconvolution of image degraded by Gaussian blur with Gaussian noise. (a) Blur estimate by CRL. (b) Blur estimate by RSD. (c) Blur estimate by SDR

(a)   (b)

(c)

Figure 7.7: Blind deconvolution of image degraded by uniform blur with additive noise. (a) Image degraded by 5×5 uniform blur with 40 dB additive noise. (b) Restored image using CRL. (c) Restored image using SDR

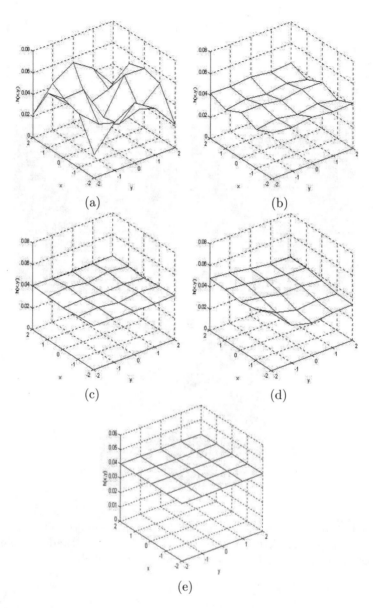

Figure 7.8: Blur estimates for blind deconvolution of image degraded by uniform blur with 40 dB additive noise. (a) Initial random blur estimate. (b) Blur estimate after a few generations. (c) Final identified blur using our approach. (d) Identified blur using SDR approach. (e) Actual 5×5 uniform blur

identification of spatially varying blurs such as Gaussian mask as shown in Figure 7.6a. The good performances of our approach illustrate that this technique is effective and robust in blind deconvolution of images degraded under different circumstances, namely, various blur structures and noise levels.

7.4.4 Identification of Non-standard Blur by RSD

We illustrate the capability of the soft-decision method to handle non-standard blur in Figure 7.9. The original image in Figure 7.2a was degraded by the 5 × 5 nonstandard exponential blur given in Figure 7.9f, followed by 30 dB additive noise to the form the degraded image in Figure 7.9a. The algorithm was applied to obtain the restored image and identified blur in Figures 7.9b and d, respectively. It is clear that this algorithm is effective in restoring the image by providing clarity in the fine texture regions, and suppressing noise and ringing in the smooth backgrounds. Compared to our approach, the restored image using SDR method in Figure 7.9c does not render enough details near the texture feather regions. Again, ringing starts to appear near the shoulder in the smooth backgrounds. An inspection on our identified blur shows that it captures the overall structure of the actual blur. In contrast, the identified blur using the SDR method in Figure 7.9e shows that SDR is ineffectual in recovering the nonstandard blurs. The satisfactory results of our approach illustrate that CRL and RSD methods are effective and robust in the blind deconvolution of images degraded under different circumstances, namely, various standard blurs, non-standard blur and different noise levels.

7.5   Conclusions

This chapter described two methods for blind image deconvolution. The first formulates blind deconvolution problem into an evolutionary strategy comprising the generation of image and blur populations. A knowledge-based stochastic initialization is developed to initialize the blur population. It utilizes the notion that the more likely search space should be assigned more emphasis. Inter-niche-space recombination is employed to introduce variation across the population spectrum, while intra-niche-space recombination strengthens the existence of solutions within similar spatial residency. A reinforced mutation scheme that combines the classical mutation and reinforced learning is developed. The scheme integrates the *a priori* information of the blur structures as dynamic attractor. This improves the convergence greatly, leading to significant reduction in computational cost.

The second applies a soft-decision blur identification and neural network approach to adaptive blind image restoration. The method incorporates the *a priori* information of well-known blur structures without compromising its flexibility in restoring images degraded by other non-standard blurs. A multimodal cost function consisting of data fidelity measure, image and blur-domain regularization terms and soft-decision estimation error is presented. A subspace

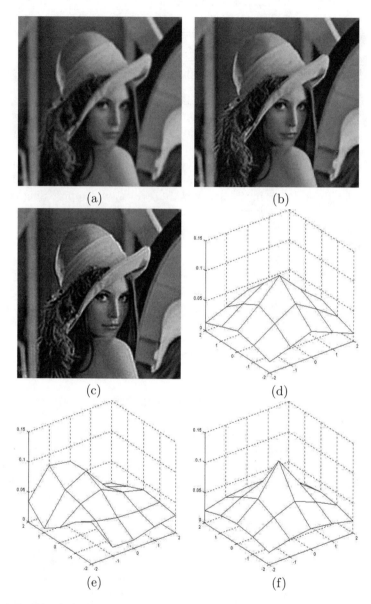

Figure 7.9: Blind deconvolution of image degraded my nonstandard blur. (a) Image degraded by 5×5 nonstandard exponential blur with 30 dB additive noise. (b) Restored image using RSD. (c) Restored image using SDR. (d) Identified blur by RSD. (e) Identified blur by SDR. (f) Actual 5×5 exponential blur

optimization process is adopted where the cost function is projected and optimized with respect to the image and blur domains recursively. The Hierarchical Cluster Model is used to achieve the perception-based restoration by minimizing the image-domain cost function. HCM has sparse synaptic connections, thereby reducing the computational cost of restoration. In addition, its unique cluster configuration is compatible with region-based human visual perception. The technique models the current blur with the best-fit parametric structure, assesses its relevance by evaluating the proximity measure and incorporates the weighted soft estimate into the blur identification criteria. It reconciles the dilemma faced by traditional blind image restoration schemes where a hard decision on the certainty of blur structure has to be made before the algorithm formulation.

Experimental results show that the methods presented in the chapter are robust in blind deconvolution of images degraded under different blur structures and noise levels. In particular, it is effective in identifying the difficult Gaussian blur that troubles most traditional blind restoration methods.

Chapter 8

Edge Detection Using Model-Based Neural Networks

8.1 Introduction

In this chapter, we adopt user-defined salient image features as training examples for a specially designed model-based neural network to perform feature detection. Specifically, we have investigated an alternative MBNN with hierarchical architecture [48, 62, 128] which is capable of learning user-specified features through an interactive process. It will be shown that this network is capable of generalizing from the set of specified features to identify similar features in images which have not been included in the training data.

Edge characterization represents an important sub-problem of feature extraction, the aim of which is to identify those image pixels with appreciable changes in intensities from their neighbors [2, 4]. The process usually consists of two stages: in the first stage, all sites with appreciable intensity changes compared to their neighbors are identified, and in the second stage, the level of intensity change associated with each site is compared against a threshold to decide if the change is "significant" enough such that the current pixel is to be regarded as an edge pixel. In other words, a suitable difference measure is required in the first stage, examples of which include the Roberts operator, the Prewitt operator and the Sobel operator [2, 4]. On the other hand, the second stage requires the specification of a threshold which describes the notion of a significant edge magnitude level.

In simple edge detection, the user specifies a global threshold on the edge magnitudes in an interactive way to produce a binary edge map from the edge magnitude map. The result is usually not satisfactory since some of the noisy pixels may be misclassified as edge pixels. More sophisticated approaches like the Canny edge detector [6] and the Shen-Castan edge detector [7] adopt the so-called hysteresis thresholding operation, where a significant edge is defined as a sequence of pixels with the edge magnitude of at least one of its members exceeding an upper threshold, and with the magnitudes of the other pixels exceeding a lower threshold. The localization of an edge within a single pixel width, in addition, requires some form of Laplacian of Gaussian (LoG) filtering [8] to detect the zero crossings, where more associated parameters have to be specified. Adding these to the previous thresholding parameter set gives rise to a large number of possible combinations of parameter values, each of which will result in a very different appearance for the edge map.

In this chapter a new MBNN architecture is adopted to estimate the set of parameters for edge characterization. In anticipation of a still larger parameter set for more sophisticated edge detection operations, a logical choice for its representation would be in the form of the set of connection weights for a neural network. To this end, an MBNN based on the hierarchical architecture proposed by Kung and Taur [48] is developed for the purpose of edge characterization, in which the connection weights play the dual role of encoding the edge-modeling parameters in the initial high-pass filtering stage and the thresholding parameters in the final decision stage.

8.2   MBNN Model for Edge Characterization

In Chapter 5, we have adopted a hierarchical model-based neural network architecture for adaptive regularization. Specifically, the network consists of a set of weight-parameterized model-based neurons as the computational units, and the output of each sub-network is defined as the linear combination of these local neuron outputs. For the current application, the MBNN architecture will again be adopted, but due to the different requirements of the edge characterization problem, an alternative neuron model, which we refer to as the input-parameterized model-based neuron, is proposed. In addition, a winner-take-all competition process is applied to all the model-based neurons within a single sub-network to determine the sub-network output instead of the previous linear combination operation.

8.2.1   Input-Parameterized Model-Based Neuron

Instead of mapping the embedded low-dimensional weight vector $\mathbf{z} \in \mathbf{R}^M$ to the high-dimensional weight vector $\mathbf{p} \in \mathbf{R}^N$ as in the case of a weight-parameterized model-based neuron, the *input-parameterized* model-based neuron is designed such that a high-dimensional *input vector* $\mathbf{x} \in \mathbf{R}^N$ is mapped to a low-dimensional vector $\mathbf{x}^P \in \mathbf{R}^M$. This is under the assumption that for the problem at hand the reduced vector $\mathbf{x}^P$ can fully represent the essential characteristics of its higher-dimensional counterpart $\mathbf{x}$. If such a mapping exists for the set of input vectors $\mathbf{x}$, we can directly represent the weight vector in its low-dimensional form $\mathbf{z}$ in the network instead of its embedded form in a high-dimensional vector $\mathbf{p}$.

More formally, we assume the existence of a mapping $\mathcal{P} : \mathbf{R}^N \longrightarrow \mathbf{R}^M$, such

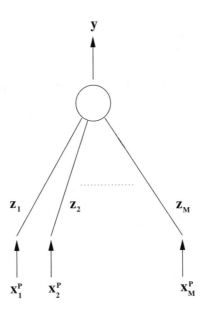

Figure 8.1: The input-parameterized model-based neuron

that $\mathbf{x}^P = \mathcal{P}(\mathbf{x}) \in \mathbf{R}^M$, where $M < N$. The operation of this model-based neuron is then defined as follows:

$$y_s = f_s(\mathbf{x}, \mathbf{p})$$
$$= f_s(\mathbf{x}^P, \mathbf{z}) \tag{8.1}$$

The input-parameterized model-based neuron is illustrated in Figure 8.1.

The two different types of model-based neurons both allow the search for the optimal weights to proceed more efficiently in a low-dimensional weight space. The decision to use either one of the above neurons depends on the nature of the problem, which in turn determines the availability of either the mapping $\mathcal{M}$ for the weights, or the mapping $\mathcal{P}$ for the input: if the parameter space of a certain problem is restricted to a low-dimensional surface in a high-dimensional space, then it is natural to adopt the weight-parameterized model-based neuron. This is the case for the adaptive regularization problem discussed previously, where the valid set of weight vectors for the network forms a path, or alternatively a one-dimensional surface, governed by the single regularization parameter λ.

On the other hand, for cases where such representation for the weight vector is not readily available, while there is evidence, possibly through the application of principal component analysis (PCA) [161, 162, 163] or prior knowledge regarding the input vector distribution, of the existence of an information-preserving operator (with respect to the current problem) which maps the input vectors to a low-dimensional subspace, then it is more natural to adopt the input-parameterized model-based neuron. This is the case for the current edge characterization model-based neural network where the original input vector comprising a window of edge pixel values is mapped to a two-dimensional vector representing the two dominant gray levels around the edge.

8.2.2   Determination of Sub-Network Output

For the current application, instead of using a linear combination approach as in the previous adaptive regularization problem, a winner-take-all competition process is applied to all the neurons within a certain sub-network to determine the *local winner* as follows:

$$\mathbf{p}_{s_r^*} = \arg\min_{s_r} \psi_r(\mathbf{x}, \mathbf{p}_{s_r}) \tag{8.2}$$

where $\mathbf{p}_{s_r^*}$ is the index of the winning neuron.

The sub-network output $\phi(\mathbf{x}, \mathbf{p}_r)$ is then substituted with the corresponding local neuron output of the winner

$$\phi(\mathbf{x}, \mathbf{p}_r) = \psi_r(\mathbf{x}, \mathbf{p}_{s_r^*}) \tag{8.3}$$

In accordance with this competition process, it is natural to adopt the Euclidean distance for evaluating the local neuron output

$$\psi_r(\mathbf{x}, \mathbf{p}_{s_r}) = \|\mathbf{x} - \mathbf{p}_{s_r}\| \tag{8.4}$$

This class of networks is especially suitable for unsupervised pattern classification, where each pattern class is composed of several disjoint subsets of slightly different characteristics. We can then assign each primary pattern class to a single sub-network, and each secondary class under the current primary class to a neuron within the sub-network. The resulting network architecture and the structure of its sub-network is shown in Figure 8.2.

For the input-parameterized model-based neuron, instead of embedding the model-based vector $\mathbf{z}_{s_r}$ in a higher-dimensional space, we map the input vector $\mathbf{x} \in \mathbf{R}^N$ on to a low-dimensional sub-manifold $\mathbf{R}^M$ through the operator $\mathcal{P}$ as follows:

$$\psi_r(\mathbf{x}, \mathbf{p}_{s_r}) \equiv \psi_r(\mathbf{x}^P, \mathbf{z}_{s_r}) \tag{8.5}$$
$$= \psi_r(\mathcal{P}(\mathbf{x}), \mathbf{z}_{s_r}) \tag{8.6}$$

where $\mathbf{x}^P = \mathcal{P}(\mathbf{x})$ is the low-dimensional input vector corresponding to $\mathbf{x}$ in $\mathbf{R}^N$.

8.2.3   Edge Characterization and Detection

The adoption of the current network architecture is motivated by our observation of the different preferences of human beings in regarding a certain magnitude of gray level discontinuity as constituting a significant edge feature under different illuminations. To incorporate this criterion into our edge detection process, it is natural to adopt a hierarchical network architecture where we designate

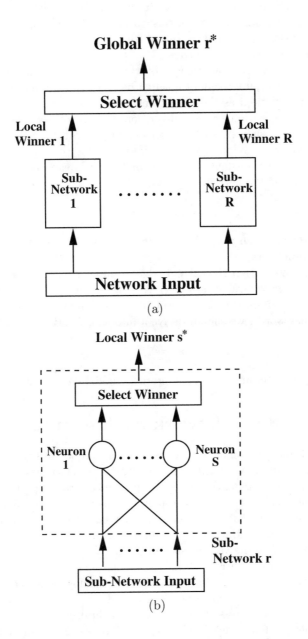

Figure 8.2: Hierarchical network architecture for edge characterization: (a) global network architecture, (b) sub-network architecture

each sub-network to represent a different illumination level, and each neuron in the sub-network to represent different prototypes of edge-like features under the corresponding illumination level. In this work, we have defined an *edge prototype* as a two-dimensional vector $\mathbf{w} \in \mathbf{R}^2$ which represents the two dominant gray level values on both sides of the edge.

Due to this necessity of inferring prototypes from the human-supplied training examples, it is natural to adopt unsupervised competitive learning where each prototype is represented as the weight vector of a neuron. The winner-take-all nature of the competition process also favors the use of the sub-cluster hierarchical architecture, such that only the local winner within each sub-network is allowed to update its weight vector. In addition, for the rth sub-network, the local neuron output is evaluated in terms of the Euclidean distance between the edge prototype and the current edge example

$$\psi_r(\mathbf{x}^P, \mathbf{z}_{s_r}) = \|\mathbf{x}^P - \mathbf{z}_{s_r}\| \tag{8.7}$$

where $\mathbf{x}^P \in \mathbf{R}^2$ is the current edge example, and $\mathbf{z}_{s_r}$ is the s_rth edge prototype of the rth sub-network.

In view of the fact that the current edge example is usually specified in terms of a window of gray level values $\mathbf{x} \in \mathbf{R}^N$, where $N \gg 2$, it is necessary to summarize this high-dimensional vector in terms of its two dominant gray level values. In other words, we should derive a mapping $\mathcal{P} : \mathbf{R}^N \longrightarrow \mathbf{R}^2$ such that

$$\mathbf{x}^P = \mathcal{P}(\mathbf{x}) \tag{8.8}$$

which corresponds to an input-parameterized model-based neuron.

8.3   Network Architecture

The proposed MBNN architecture consists of a number of sub-networks, with each neuron of a sub-network encoding a specified subset of the training samples. In the initial training stage, the edge examples in the training set are adaptively partitioned into different subsets through an unsupervised competitive process between the sub-networks. In the subsequent recognition stage, edge pixels are identified by comparing the current pixel configuration with the encoded templates associated with the various sub-networks.

The adoption of the current HMBNN architecture is due to our observation that, in edge detection, it would be more effective to adopt multiple sets of thresholding decision parameters corresponding to different local contexts, instead of a single parameter set across the whole image as in previous approaches.

We consider the following encoding scheme where each sub-network is associated with an edge template corresponding to a different background illumination level, and each neuron in the sub-network encodes possible variations of edge prototypes under the corresponding illumination level. The architecture of the feature detection network is shown in Figure 8.3 and the hierarchical encoding scheme for the edge prototypes are described in the following sections.

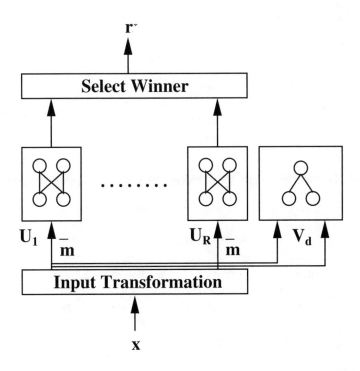

Figure 8.3: The architecture of the HMBNN edge detector

8.3.1   Characterization of Edge Information

The edge detection is performed on the $N \times N$ neighborhood of the current pixel. Concatenating the corresponding gray level values into a vector $\mathbf{x} = [x_1 \ \ldots \ x_{N^2}]^T \in \mathbf{R}^N$, the mean of the gray is evaluated as follows:

$$\overline{x} = \frac{1}{N^2} \sum_{n=1}^{N^2} x_n \qquad (8.9)$$

Given this mean value, the gray level information in this local window can be further characterized by a vector $\mathbf{m} = [m_1 \ m_2]^T \in \mathbf{R}^2$, the components of which correspond to the two dominant gray level values within the window and is defined as follows:

$$m_1 = \frac{\sum_{i=1}^{N^2} I(x_i < \overline{x}) x_i}{\sum_{i=1}^{N^2} I(x_i < \overline{x})} \qquad (8.10)$$

$$m_2 = \frac{\sum_{i=1}^{N^2} I(x_i \geq \overline{x}) x_i}{\sum_{i=1}^{N^2} I(x_i \geq \overline{x})} \qquad (8.11)$$

$$\overline{m} = \frac{m_1 + m_2}{2} \qquad (8.12)$$

where the function $I(\cdot)$ equals 1 if the argument is true, and equals 0 otherwise.

8.3.2   Sub-Network U_r

As described previously, each sub-network $U_r, r = 1, \ldots, R$ is associated with a prototype background illumination gray level value p_r. Those local $N \times N$ windows in the image with their mean values closest to p_r are then assigned to the sub-network U_r for further encoding. Specifically, a particular pixel window $\mathcal{W}$ in the image, with its associated mean values $\overline{m}, m_1, m_2$ as defined in equations (8.10) to (8.12), is assigned to the sub-network U_{r^*} if the following conditions are satisfied

$$p_{r^*} \in [m_1, m_2] \qquad (8.13)$$

$$|\overline{m} - p_{r^*}| < |\overline{m} - p_r| \qquad r = 1, \ldots, R, r \neq r^* \qquad (8.14)$$

where $[m_1, m_2]$ is the closed interval with m_1, m_2 as its endpoints. The set of all $N \times N$ pixel windows are thus partitioned into subsets with the members of each subset exhibiting similar levels of background illumination. For convenience, we will refer to the two conditions in (8.13) and (8.14) collectively as $\mathbf{x} \longrightarrow U_{r^*}$. The architecture of the sub-network is shown in Figure 8.4.

8.3.3   Neuron V_{rs} in Sub-Network U_r

Each sub-network U_r consists of S neurons $V_{rs}, s = 1, \ldots, S$, which encode the local edge templates representing the possible variations under the general illumination level p_r. Each neuron is associated with a weight vector $\mathbf{w}_{rs} =$

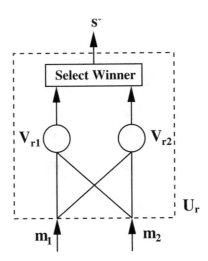

Figure 8.4: The architecture of a single sub-network U_r

$[w_{rs,1} \ w_{rs,2}]^T \in \mathbf{R}^2$ which summarizes the two dominant gray level values in each N × N window $\mathcal{W}$ in the form of a local prototype vector $\mathbf{m}$. A window with an associated vector $\mathbf{m}$ is assigned $V_{r^*s^*}$ if the following condition holds

$$\|\mathbf{m} - \mathbf{w}_{r^*s^*}\| < \|\mathbf{m} - \mathbf{w}_{rs}\| \qquad s = 1, \ldots, S, s \neq s^* \qquad (8.15)$$

We have chosen $S = 2$, i.e., each sub-network consists of two neurons, in order that one of the neurons encodes the local prototype for weakly visible edges and the other encodes the prototype for strongly visible edges. Correspondingly, one of the weight vectors $\mathbf{w}_{r^*s}, s = 1, 2$ is referred to as the weak edge prototype $\mathbf{w}_{r^*}^l$ and the other one as the strong edge prototype $\mathbf{w}_{r^*}^u$. The determination of whether a neuron corresponds to a strong or weak edge prototype is made according to the following criteria:

- $\mathbf{w}_{r^*}^l = \mathbf{w}_{rs'}$, where $s' = \arg\min_s (w_{r^*s,2} - w_{r^*s,1})$

- $\mathbf{w}_{r^*}^u = \mathbf{w}_{rs''}$, where $s'' = \arg\max_s (w_{r^*s,2} - w_{r^*s,1})$

Given the weak edge prototype $\mathbf{w}_{r^*}^l = [w_{r^*,1}^l \ w_{r^*,2}^l]^T$, the measure $(w_{r^*,2}^l - w_{r^*,1}^l)$ plays a similar role as the threshold parameter in conventional edge detection algorithms in specifying the lower limit of visibility for edges, and is useful for identifying potential starting points in the image for edge tracing. The structure of the neuron is shown in Figure 8.5.

8.3.4 Dynamic Tracking Neuron V_d

In addition to the sub-networks U_r and the local neurons V_{rs}, a *dynamic tracking neuron* V_d with the network itself. In other words, this neuron is global in scope

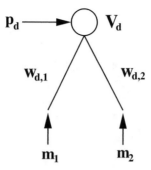

Figure 8.5: The structure of the dynamic edge tracking neuron

and does not belong to any of the sub-networks U_r. The dynamic neuron is a hybrid between a sub-network and a local neuron in that it consists of *both* a *dynamic weight vector* $\mathbf{w}_d = [w_{d,1} \quad w_{d,2}]^T \in \mathbf{R}^2$, which corresponds to the weight vector of the local neuron, and a scalar parameter p_d which is analogous to the illumination level indicator p_r of each sub-network. The structure of the dynamic tracking neuron is shown in Figure 8.5.

The purpose of this neuron is to track the varying background gray levels during the edge tracking operation. The neuron is inactive in the training stage. In the recognition stage and upon the detection of a prominent edge point, the dynamic weight vector $\mathbf{w}_d$ and illumination level indicator p_d are continuously modified to track those less prominent edge points connected to the initial edge point.

8.3.5   Binary Edge Configuration

We suppose that the vector $\mathbf{m}$ for the current window $\mathcal{W}$ is assigned to neuron $V_{r^*s^*}$ with weight vector $\mathbf{w}_{r^*s^*}$. For edge detection, the real-valued vector $\mathbf{x} \in \mathbf{R}^{N^2}$ representing the gray level values of the current window is mapped to a binary vector $\mathbf{b} \in \mathbf{B}^{N^2}$, where $\mathbf{B} = \{0,1\}$. To achieve this purpose, we define the mapping $\mathcal{Q} : \mathbf{R}^{N^2} \times \mathbf{R}^2 \longrightarrow \mathbf{B}^{N^2}$ as follows:

$$\mathbf{b} = \mathcal{Q}(\mathbf{x}, \mathbf{w}_{r^*s^*}) = [q(x_1, \mathbf{w}_{r^*s^*}) \quad \ldots \quad q(x_{N^2}, \mathbf{w}_{r^*s^*})]^T \in \mathbf{B}^{N^2} \qquad (8.16)$$

where the component mappings $q : \mathbf{R} \times \mathbf{R}^2 \longrightarrow \mathbf{B}$ are specified as

$$q(x_n, \mathbf{w}_{r^*s^*}) = \begin{cases} 0 & \text{if } |x_n - w_{r^*s^*,1}| < |x_n - w_{r^*s^*,2}| \\ 1 & \text{if } |x_n - w_{r^*s^*,1}| \geq |x_n - w_{r^*s^*,2}| \end{cases} \qquad (8.17)$$

For valid edge configurations, the binary vectors $\mathbf{b}$ assume special forms which are illustrated in Figure 8.6 for $N = 3$.

During the initial training stage, the binarized edge configuration associated with each training edge pattern is stored in an *edge configuration set* C which

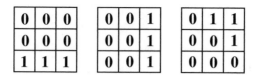

Figure 8.6: Examples of valid edge configurations

Figure 8.7: Illustrating the operation $R_{\frac{\pi}{4}}$

forms part of the overall network parameter set. The set C is then expanded in order that, when $N = 3$, it is closed under an operation $R_{\frac{\pi}{4}}$. Specifically, this operation permutes the entries of a vector $\mathbf{b} \in \mathbf{B}^9$ in such a way that, when interpreted as the entries in a 3×3 window, $R_{\frac{\pi}{4}}(\mathbf{b})$ is the 45° clockwise rotated version of $\mathbf{b}$. This is illustrated in Figure 8.7. In this way, detection of those rotated edge configurations not present in the training set is facilitated.

8.3.6 Correspondence with the General HMBNN Architecture

The current edge characterization network has a direct correspondence with the general MBNN architecture as follows: the sub-networks $U_r, r = 1, \ldots, R$ can be directly associated with the sub-networks in the general model, and the neurons $V_{rs_r}, s_r = 1, \ldots, S_r$ are associated with those within each sub-network in the general model, with $S_r = 2$ for all r in the current application.

Instead of embedding a low-dimensional weight vector in a high-dimen-sional space as with the weight-parameterized model-based neuron, we have instead mapped the relatively high-dimensional input vector $\mathbf{x}$ into a low-dimensional input $\mathbf{m}$ which characterizes the two dominant gray level values within a local window of pixels. In other words, we adopt the input-parameterized model-based neuron described previously, with the mapping $\mathcal{P} : \mathbf{R}^{N^2} \longrightarrow \mathbf{R}^2$, defined such that

$$\mathbf{m} = \mathcal{P}(\mathbf{x}) \tag{8.18}$$

Despite the above correspondences, there are slight differences between the original formulation of the subcluster hierarchical structure in [48] and the current edge characterization network structure. In the original network, the sub-network output $\phi(\mathbf{x}, \mathbf{w_r})$ is directly substituted by the neuron output of the local winner as follows:

$$\phi(\mathbf{x}, \mathbf{w}_r) \equiv \psi_r(\mathbf{x}, \mathbf{w}_{s^*}) \qquad (8.19)$$

where s^* is the index of the local winner, and $\psi_r(\mathbf{x}, \mathbf{w}_{s^*})$ is the local neuron output of the winning neuron.

In the current network, on the other hand, the competition process at the sub-network level is independent of the corresponding process at the neuronal level, and is of a very different nature: the competition between the sub-networks is specified in terms of the conformance of the current local illumination gray level value with one of the prototype background illumination levels, i.e., a comparison between scalar values. The competition at the neuronal level, however, involves the comparison of vectors in the form of two-dimensional edge prototypes existing under a particular illumination gray level value. As a result, the sub-network outputs and the local neuron outputs are independently specified at the two hierarchical levels. At the sub-network level, the following sub-network output is defined for the background illumination gray level values:

$$\phi(\overline{m}, p_r) = |\overline{m} - p_r| \qquad (8.20)$$

At the neuron level, the following local neuron output is defined for the edge prototype vectors:

$$\psi_r(\mathbf{m}, \mathbf{w}_{rs}) = \|\mathbf{m} - \mathbf{w}_{rs}\| \qquad (8.21)$$

8.4   Training Stage

The training stage consists of three sub-stages: in the first sub-stage, the prototype background gray level level $p_r, r = 1, \ldots, R$ for each sub-network U_r is determined by competitive learning [34, 44]. In the second sub-stage, each window $\mathcal{W}$ is assigned to a sub-network and a local neuron under this sub-network based on its associated parameters p_r and $\mathbf{m}$. The weight vector $\mathbf{w}_{rs}$ of the winning local neuron is then updated using competitive learning. In the third stage, the corresponding binary edge configuration pattern $\mathbf{b}$ is extracted as a function of the winning weight vector $\mathbf{w}_{rs}$ based on the mapping $\mathcal{Q}$. Adopting a window size of $N = 3$, we apply the operation $R_{\frac{\pi}{4}}$ successively to $\mathbf{b}$ to obtain the eight rotated versions of this pattern, and insert these patterns into the edge configuration memory C.

8.4.1   Determination of p_{r^*} for Sub-Network U_{r^*}

Assuming that the current window with associated mean p_r is assigned to U_{r^*} based on conditions (8.13) and (8.14). The value of p_{r^*} is then updated using competitive learning as follows:

$$p_{r^*}(t + 1) = p_{r^*}(t) + \eta(t)(\overline{m} - p_{r^*}(t)) \qquad (8.22)$$

The learning stepsize $\eta(t)$ is successively decreased according to the following schedule:

$$\eta(t+1) = \eta(0)\left(1 - \frac{t}{t_f}\right) \tag{8.23}$$

where t_f is the total number of iterations.

8.4.2 Determination of $\mathbf{w}_{r^*s^*}$ for Neuron $V_{r^*s^*}$

Assuming that the current window with associated feature vector $\mathbf{m}$ is assigned to the local neuron $V_{r^*s^*}$ under sub-network U_{r^*}, the associated weight vector $\mathbf{w}_{r^*s^*}$ is again updated through competitive learning as follows:

$$\mathbf{w}_{r^*s^*}(t+1) = \mathbf{w}_{r^*s^*}(t) + \eta(t)(\mathbf{m} - \mathbf{w}_{r^*s^*}(t)) \tag{8.24}$$

where the stepsize $\eta(t)$ is successively decreased according to equation (8.23).

8.4.3 Acquisition of Valid Edge Configurations

After going through the previous stages, the background illumination level indicators p_r for the sub-networks and the weight vectors $\mathbf{w}_{rs}$ for the neurons have all been determined. As a result, all the $N \times N$ windows can be assigned to their corresponding sub-networks and the correct neurons within the sub-networks according to their parameters $\overline{m}$ and $\mathbf{m}$. If the current window is assigned to neuron $V_{r^*s^*}$ under sub-network U_{r^*}, the gray level vector $\mathbf{x}$ of the current window $\mathcal{W}$ can be transformed into a binary edge configuration vector $\mathbf{b}$ as a function of $\mathbf{w}_{r^*s^*}$ according to equation (8.17):

$$\mathbf{b} = \mathcal{Q}(\mathbf{x}, \mathbf{w}_{r^*s^*}) \tag{8.25}$$

The binary vector $\mathbf{b}$ is then inserted into the valid edge configuration set C. Using a window size of $N = 3$, the requirement that the set C be closed under the operation $R_{\frac{\pi}{4}}$ can be satisfied by generating the following eight edge configurations $\mathbf{b}_j, j = 0, \ldots, 7$ using $R_{\frac{\pi}{4}}$ as follows

$$\mathbf{b}_0 = \mathbf{b} \tag{8.26}$$
$$\mathbf{b}_{j+1} = R_{\frac{\pi}{4}}(\mathbf{b}_j) \quad j = 0, \ldots, 6 \tag{8.27}$$

and inserting all of them into the configuration set C.

8.5 Recognition Stage

In this stage, all pixels in a test image are examined in order that those pixels with edge-like features are identified. This recognition stage consists of two sub-stages. In the first sub-stage, all pixels in the test image with high degree of similarity to the learned edge prototypes are declared as primary edge points. In the second sub-stage, these primary edge points are used as starting points for an edge tracing operation such that the less prominent edge points, which we refer to as secondary edge points, are recursively identified.

8.5.1   Identification of Primary Edge Points

In this sub-stage, all $N \times N$ windows in the test image are examined. All those windows with associated parameters $\overline{m}, m_1, m_2$ satisfying the following conditions are declared as primary edge points.

(A1). $\mathbf{x} \longrightarrow U_{r^*}$ (i.e., satisfaction of conditions (8.13) and (8.14)) for some r^*.

(A2). $m_2 - m_1 \geq w^l_{r^*,2} - w^l_{r^*,1}$, where $\mathbf{w}^l_{r^*}$ is the weak edge prototype vector of U_{r^*}.

(A3). $\mathbf{b} = \mathcal{Q}(\mathbf{x}, \mathbf{w}_{r^*s^*}) \in C$, where $\mathbf{w}_{r^*s^*}$ is the weight vector associated with the selected neuron $V_{r^*s^*}$.

Condition (A1) specifies that the mean gray level value of the current window should be close to one of the designated levels of the network. Condition (A2) ensures that the edge magnitude, as represented by the difference $m_2 - m_1$, is greater than the difference between the components of the corresponding weak edge prototype. Condition (A3) ensures that the binary edge configuration corresponding to the current window is one of the valid templates in the configuration set C.

8.5.2   Identification of Secondary Edge Points

In this second sub-stage, the dynamic tracking neuron V_d is activated to trace the secondary edge points connected to the current primary edge point. The gray level indicator p_d and weight vector $\mathbf{w}_d$ of the neuron are initialized using the parameters of the detected primary edge point $\overline{m}_p$ and $\mathbf{m}^p = [m^p_1 \ m^p_2]^T$ as follows:

$$p_d(0) = \overline{m}^p \tag{8.28}$$
$$\mathbf{w}_d(0) = \mathbf{m}^p \tag{8.29}$$

After initializing the parameters of the dynamic neuron, a recursive edge tracing algorithm is applied to identify the less prominent edge pixels (the secondary edge points) connected to the primary edge points by applying the following set of conditions at each 8-neighbor of the current primary edge point:

(B1). $\mathbf{b} = \mathcal{Q}(\mathbf{x}, \mathbf{w}_d) \in C$.

(B2). $p_d \in [m_1, m_2]$.

Condition (B1) is similar to condition (A3) for the primary edge point detection, while condition (B2) is a modified version of condition (8.13) which forms part of the requirements for $\mathbf{x} \longrightarrow U_r$, to ensure that the mean gray level value of a potential secondary edge point, as represented by p_d, should be similar to those of the previously traversed edge points. In addition, no specific conditions are imposed on the edge magnitude of the current pixel to allow the possibility of including weak edge points in the final edge segment, as long as it is connected to a prominent edge point.

For each secondary edge point (with corresponding parameters $\overline{m}^s$ and $\mathbf{m}^s$) which satisfies the above conditions, the local illumination level indicator p_d of the dynamic neuron is updated as follows:

$$p_d(t+1) = p_d(t) + \eta(t)(\overline{m}^s - p_d(t)). \qquad (8.30)$$

In addition, if this edge point satisfies condition (A2) for primary edge point detection, indicating that its edge magnitude is comparable to that of a primary edge point, the weight vector $\mathbf{w}_d$ of the dynamic neuron is also updated to incorporate the characteristics of the current point:

$$\mathbf{w}_d(t+1) = \mathbf{w}_d(t) + \eta(t)(\mathbf{m}^s - \mathbf{w}_d(t)) \qquad (8.31)$$

8.6 Experimental Results

In the training stage, users are requested to select different edge examples from two images depicting a flower and an eagle, respectively. These are then used as training examples for the network. To ensure that the users' preferences are adequately represented, different edge traces have been prepared for each of the two images. The eagle image and three different sets of edge examples are shown in Figure 8.8, while those for the flower image are shown in Figure 8.9.

After the training stage, the MBNN is first applied to images in the training set in order to test the capability of the network to identify all relevant edge-like features from the previous sparse tracings. The result is shown in Figure 8.10a.

By comparing this figure with the tracings in Figures 8.8b to d, we can conclude that the network is capable of generalizing from the initial sparse tracings to identify the important edges. In addition, we compare this result with that of a standard edge detector. Since the performance of standard edge detectors depends on the values of a set of tunable parameters, our result can be validated if the performance of our approach is comparable to the standard results under near-optimal parameter settings.

The Shen-Castan edge detector [7] is chosen as the comparison standard in Figure 8.10. This edge detector employs an exponential filter in the smoothing operation prior to the differencing operation, which provides a better degree of edge localization than the Gaussian filter used in the Canny edge detector [6]. For this edge detector, the hysteresis thresholds t_1, t_2 are adjusted in each edge detection operation to provide the best visual result.

In Figure 8.10, the performance of the MBNN edge detector is compared with that of the Shen-Castan edge detector under different hysteresis threshold settings. The MBNN result is shown in Figure 8.10a and the Shen-Castan edge detection results are shown in Figures 8.10b to d. The lower hysteresis threshold ranges from $t_1 = 10$ to $t_1 = 40$, and the upper threshold is set to $t_2 = t_1 + 5$.

We can observe from Figures 8.10b to d that the set of detected edges are sensitive to the choice of t_1 and t_2: lower values of t_1 and t_2 reveal more details but result in more false positive detections as seen in Figure 8.10b, while higher threshold values lead to missed features as in Figure 8.10d. In our opinion, Figure 8.10c, with $t_1 = 20$ and $t_2 = 25$, constitutes an acceptable representation

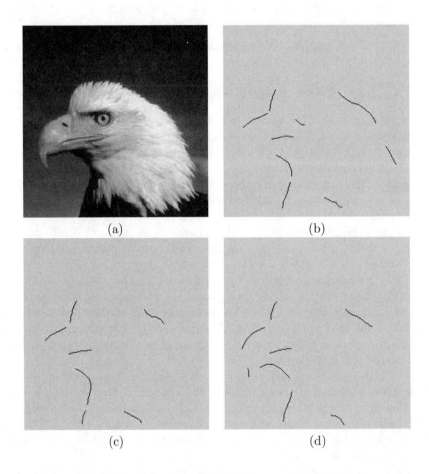

Figure 8.8: (a) Eagle image. (b)-(d) User-specified edge examples

(a)

(b)

(c)

(d)

Figure 8.9: (a) Flower image. (b)-(d) User-specified edge examples

Figure 8.10: Edge detection results for the eagle image: (a) Detected edges using NN, (b)-(d) Detected edges using Shen-Castan edge detector with different hysteresis thresholds t_1, t_2: (b) $t_1 = 10, t_2 = 15$; (c) $t_1 = 20, t_2 = 25$; (d) $t_1 = 30, t_2 = 35$

of the underlying edge features. These can be compared with the MBNN result in Figure 8.10a, where we can see that the edge map is similar to that of Figure 8.10c which we have previously chosen as the one corresponding to the preferred threshold parameter settings for the Shen-Castan edge detector.

The results for the flower image are shown in Figure 8.11. Figures 8.11b to d show the effect of varying the thresholds t_1, t_2 and the MBNN result is shown in Figure 8.11a. For the Shen-Castan edge detection results, we may consider Figure 8.11c (with thresholds $t_1 = 20$ and $t_2 = 25$) as a near optimal representation and Figure 8.11d as an adequate representation, although there are some missing features in this case. In Figure 8.11a, we notice that the MBNN detection result lies in between these two and thus can be considered a close approximation to the optimal result. On the other hand, we can see that the result in Figure 8.11b is over-cluttered with non-essential details due to the low threshold values.

The generalization performance of the NN edge detector is further evaluated by applying the network trained on the eagle and flower images to previous unseen images. In Figure 8.12a, image of a building is shown, and Figure 8.12c shows an image of a plane. The corresponding detection results are shown in Figures 8.12b and d, respectively. The results are very satisfactory, considering that the network is trained only on the eagle and flower images.

To test the robustness of the current approach, we apply the network to noise-contaminated images. Figure 8.13a shows the result of adding zero-mean Gaussian noise with a standard deviation of $\sigma_n = 10$ to the eagle image. The previous MBNN (trained using the noiseless images) is applied to this noisy image without any re-training and alteration of architecture. The result is shown in Figure 8.13b, showing that although some false alarms occurred, the overall effect is not serious and the result is reasonably similar to the noiseless case. On the other hand, for the Shen-Castan edge detector, if we choose the previous optimal threshold of $t_1 = 20$ and $t_2 = 25$ (Figure 8.13c), the effect of noise is clearly noticeable, and the thresholds have to be re-adjusted to $t_1 = 25$ and $t_2 = 30$ to remove the noise specks (Figure 8.13d).

For the other three images, the results for the noisy case are shown in Figures 8.14b, d and f. We can again notice that the effect of noise contamination is not significant.

8.7 Summary

We have developed a model-based feature detection neural network with hierarchical architecture (HMBNN) which directly learns the essential characteristics of user-specified features through a training process. The specific architecture of the network allows the division of the training edge examples into sub-classes such that the different preferences of users in regarding intensity discontinuities as edges under different illumination conditions are taken into account. To achieve this, the MBNN implicitly represents these required parameters for edge detection in the form of network weights which are updated during the training

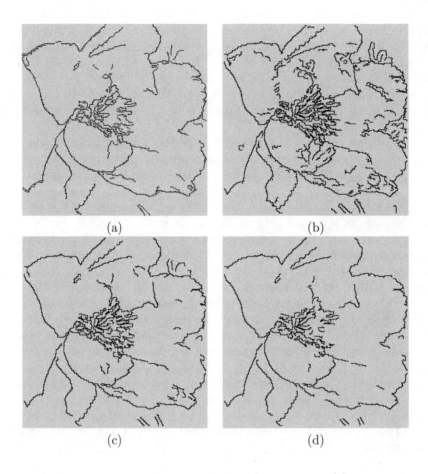

Figure 8.11: Edge detection results for the flower image. (a) Detected edges using NN. (b)-(d) Detected edges using Shen-Castan edge detector with different hysteresis thresholds t_1, t_2: (b) $t_1 = 10, t_2 = 15$; (c) $t_1 = 20, t_2 = 25$; (d) $t_1 = 30, t_2 = 35$

Figure 8.12: (a) Image of building. (b) Detected edges using NN. (c) Image of plane. (d) Detected edges using NN

Figure 8.13: (a) Eagle image with additive Gaussian noise ($\sigma_n = 10$). (b) Detected edges using NN. (c)-(d) Detected edges using Shen-Castan edge detector with different hysteresis thresholds t_1, t_2: (c)$t_1 = 20, t_2 = 25$; (d) $t_1 = 25, t_2 = 30$

(a)

(b)

(c)

(d)

(e)

(f)

Figure 8.14: (a) Flower image with additive Gaussian noise ($\sigma_n = 10$). (b) Detected edges using NN. (c) Building image with additive Gaussian noise ($\sigma_n = 10$). (d) Detected edges using NN. (e) Plane image with additive Gaussian noise ($\sigma_n = 10$). (f) Detected edges using NN

process. The current approach also takes into account of the local variations
in intensity distributions along edge segments, and the rules which capture this
non-stationarity are learned by the proposed architecture. This HMBNN edge
detector has been successfully applied to both the set of training images and to
previous unseen images with promising results. In addition, no re-training of the
network and no alteration of architecture are required for applying the network
to noisy images.

References

[1] A. K. Jain, "Advances in mathematical models for image processing," *Proc. IEEE*, vol. 69, no. 5, pp. 502–528, 1981.

[2] A. K. Jain, *Fundamentals of Digital Image Processing*. Englewood Cliffs, NJ: Prentice Hall, 1989.

[3] A. Rosenfeld, ed., *Image Modeling*. New York: Academic Press, 1981.

[4] R. C. Gonzalez and R. Woods, *Digital Image Processing*. Reading, MA: Addison-Wesley, 1992.

[5] R. M. Haralick, "Digital step edges from zero crossings of second directional derivatives," *IEEE Trans. Pattern Analysis and Machine Intelligence*, vol. 6, no. 1, pp. 58–68, 1984.

[6] J. Canny, "A computational approach to edge detection," *IEEE Trans. Pattern Analysis and Machine Intelligence*, vol. 8, no. 6, pp. 679–698, 1986.

[7] J. J. Shen and S. S. Castan, "An optimal linear operator for step edge detection," *CVGIP: Graphical Models and Image Processing*, vol. 54, no. 2, pp. 112–133, 1992.

[8] D. Marr and E. Hildreth, "Theory of edge detection," *Proc. Royal Soc.*, vol. B-207, pp. 187–217, 1980.

[9] P. Perona and J. Malik, "Scale-space and edge detection using anisotropic diffusion," *IEEE Trans. Pattern Analysis and Machine Intelligence*, vol. 12, no. 7, pp. 629–639, 1990.

[10] R. Chellappa and S. Chatterjee, "Classification of textures using Gaussian Markov random fields," *IEEE Trans. Acoust., Speech, Signal Processing*, vol. 33, no. 4, pp. 959–963, 1985.

[11] R. Chellappa and R. L. Kashyap, "Digital image restoration using spatial interaction models," *IEEE Trans. Acoust., Speech, Signal Processing*, vol. 30, no. 6, pp. 461–472, 1982.

[12] G. R. Cross and A. K. Jain, "Markov random field texture models," *IEEE Trans. Pattern Analysis and Machine Intelligence*, vol. 5, no. 1, pp. 25–39, 1983.

[13] R. L. Kashyap and R. Chellappa, "Estimation and choice of neighbors in spatial interaction models of images," *IEEE Trans. Inform. Theory*, vol. 29, no. 1, pp. 60–72, 1983.

[14] J. W. Woods, "Two-dimensional discrete Markovian fields," *IEEE Trans. Inform. Theory*, vol. 18, no. 2, pp. 232–240, 1972.

[15] H. Derin and H. Elliott, "Modelling and segmentation of noisy and textured images using Gibbs random fields," *IEEE Trans. Pattern Analysis and Machine Intelligence*, vol. 9, no. 1, pp. 39–55, 1987.

[16] S. Geman and D. Geman, "Stochastic relaxation, Gibbs distribution, and the Bayesian restoration of images," *IEEE Trans. Pattern Analysis and Machine Intelligence*, vol. 6, no. 6, pp. 721–741, 1984.

[17] M. Barnsley, *Fractals Everywhere*. New York: Academic Press, 1988.

[18] B. Mandelbrot, *Fractals and a New Geometry of Nature*. San Francisco, CA: Freeman, 1981.

[19] G. Arce and R. Foster, "Detail-preserving ranked-order based filters for image processing," *IEEE Trans. Acoust., Speech, Signal Processing*, vol. 37, no. 1, pp. 83–98, 1989.

[20] A. Tekalp, H. Kaufman, and J. Woods, "Edge-adaptive Kalman filtering for image restoration with ringing suppression," *IEEE Trans. Acoust., Speech, Signal Processing*, vol. 37, no. 6, pp. 892–899, 1989.

[21] H. Andrews and B. Hunt, *Digital Image Restoration*. Englewood Cliffs, NJ: Prentice Hall, 1977.

[22] M. R. Banham and A. K. Katsaggelos, "Digital image restoration," *IEEE Signal Processing Magazine*, vol. 14, no. 2, pp. 24–41, 1997.

[23] A. K. Katsaggelos, ed., *Digital Image Restoration*. Berlin, Germany: Springer Verlag, 1991.

[24] M. Bertero, T. Poggio, and V. Torre, "Ill-posed problems in early vision," *Proc. IEEE*, vol. 76, no. 8, pp. 869–889, 1988.

[25] B. R. Hunt, "The application of constrained least squares estimation to image restoration by digital computers," *IEEE Trans. Comput.*, vol. 22, no. 9, pp. 805–812, 1973.

[26] T. Poggio, V. Torre, and C. Koch, "Computational vision and regularization theory," *Nature*, vol. 317, no. 6035, pp. 314–319, 1985.

[27] A. K. Katsaggelos and M. G. Kang, "Spatially adaptive iterative algorithm for the restoration of astronomical images," *Int. J. of Imaging Systems and Technology*, vol. 6, no. 4, pp. 305–313, 1995.

[28] R. L. Lagendijk, J. Biemond, and D. E. Boekee, "Regularized iterative image restoration with ringing reduction," *IEEE Trans. Acoust., Speech, Signal Processing*, vol. 36, no. 12, pp. 1874–1888, 1988.

[29] H. S. Wong and L. Guan, "Adaptive regularization in image restoration using a model-based neural network," *Opt. Eng.*, vol. 36, no. 12, pp. 3297–3308, 1997.

[30] N. Ansari and E. Hou, *Computational Intelligence for Optimization*. Boston: Kluwer Academic, 1997.

[31] W. Pedrycz, *Computational Intelligence: an Introduction*. Boca Raton, FL: CRC Press, 1997.

[32] P. H. Winston, *Artificial Intelligence*. Reading, MA.: Addison-Wesley, 1984.

[33] J. S. R. Jang, C. T. Sun, and E. Mizutani, *Neuro-Fuzzy and Soft Computing: a Computational Approach to Learning and Machine Intelligence*. Upper Saddle River, NJ: Prentice Hall, 1997.

[34] S. Haykin, *Neural Networks: A Comprehensive Foundation*. New York: Macmillan, 1994.

[35] J. Hertz, A. Krogh, and R. G. Palmer, *Introduction to the Theory of Neural Computation*. Santa Fe Institute, Addison-Wesley, 1991.

[36] G. J. Klir and T. A. Folger, *Fuzzy Sets, Uncertainty and Information*. Englewood Cliffs, NJ: Prentice Hall, 1988.

[37] B. Kosko, *Neural Networks and Fuzzy Systems*. Englewood Cliffs, NJ: Prentice Hall, 1992.

[38] L. A. Zadeh, "Fuzzy sets," *Information and Control*, vol. 8, no. 3, pp. 338–353, 1965.

[39] T. Bäck, *Evolutionary Algorithms in Theory and Practice*. New York: Oxford University Press, 1996.

[40] D. B. Fogel, *Evolutionary Computation: Toward a New Philosophy of Machine Intelligence*. Piscataway, NJ: IEEE Press, 1995.

[41] E. Ardizzone, A. Chella, and R. Pirrone, "A neural based approach to image segmentation," in *Proc. Int. Conf. on Artificial Neural Networks (ICANN'94)*, pp. 1153–1156, 1994.

[42] A. Ghosh, N. R. Pal, and S. K. Pal, "Image segmentation using a neural network," *Biological Cybernetics*, vol. 66, no. 2, pp. 151–158, 1991.

[43] C. L. Huang, "Parallel image segmentation using modified Hopfield network," *Pattern Recognition Letters*, vol. 13, no. 5, pp. 345–353, 1992.

[44] T. Kohonen, *Self-Organizing Maps*. Berlin: Springer-Verlag, 2nd ed., 1997.

[45] C. Peterson and B. Söderberg, "A new method for mapping optimization problems onto neural networks," *Int. J. Neural Systems*, vol. 1, no. 1, pp. 3–22, 1989.

[46] Y. T. Zhou, R. Chellappa, A. Vaid, and B. K. Jenkins, "Image restoration using a neural network," *IEEE Trans. Acoust., Speech, Signal Processing*, vol. 36, no. 7, pp. 1141–1151, 1988.

[47] R. Anand, K. Mehrotra, C. K. Mohan, and S. Ranka, "Efficient classification for multiclass problems using modular neural networks," *IEEE Trans. Neural Networks*, vol. 6, no. 1, pp. 117–124, 1995.

[48] S. Y. Kung and J. S. Taur, "Decision-based neural networks with signal/image classification applications," *IEEE Trans. Neural Networks*, vol. 6, no. 1, pp. 170–181, 1995.

[49] E. R. Kandel and J. H. Schwartz, *Principles of Neural Science*. New York: Elsevier, 1985.

[50] R. F. Thompson, *An Introduction to Neuroscience*. New York: W.H. Freeman & Company, 1985.

[51] Y. S. Choi and R. Krishnapuram, "A robust approach to image enhancement based on fuzzy logic," *IEEE Trans. Image Processing*, vol. 6, no. 6, pp. 808–825, 1997.

[52] N. Fang and M. C. Cheng, "An automatic crossover point selection technique for image enhancement using fuzzy sets," *Pattern Recognition Letters*, vol. 14, no. 5, pp. 397–406, 1993.

[53] J. Hsieh, "Image enhancement with a fuzzy logic approach," *Electronic Letters*, vol. 31, no. 9, pp. 708–710, 1995.

[54] R. Krishnapuram, J. M. Keller, and Y. Ma, "Quantitative analysis of properties and spatial relations of fuzzy image regions," *IEEE Trans. Fuzzy Systems*, vol. 1, no. 3, pp. 222–233, 1993.

[55] C. Y. Tyan and P. P. Wang, "Image processing-enhancement, filtering and edge detection using the fuzzy logic approach," in *2nd IEEE Int. Conf. on Fuzzy Systems*, pp. 600–605, 1993.

[56] J. C. Bezdek, *Pattern Recognition with Fuzzy Objective Function Algorithms*. New York: Plenum Press, 1981.

[57] R. J. Hathaway and J. C. Bezdek, "Local convergence of the fuzzy c-means algorithms," *Pattern Recognition*, vol. 19, no. 6, pp. 477–480, 1986.

[58] N. R. Pal and J. C. Bezdek, "On cluster validity for the fuzzy c-means model," *IEEE Trans. Fuzzy Systems*, vol. 3, no. 3, pp. 370–379, 1995.

[59] M. J. Sabin, "Convergence and consistency of fuzzy c-means/ISODATA algorithms," *IEEE Trans. Pattern Analysis and Machine Intelligence*, vol. 9, no. 5, pp. 661–668, 1987.

[60] J. MacQueen, "Some methods for classification and analysis of multivariate observations," in *Proc. 5th Berkeley Symp. on Math. Stat. and Prob.*, vol. 1, pp. 281–296, 1967.

[61] D. E. Goldberg, *Genetic Algorithms in Search, Optimization and Machine Learning*. Reading, MA: Addison-Wesley, 1989.

[62] T. Caelli, D. Squire, and T. Wild, "Model-based neural networks," *Neural Networks*, vol. 6, no. 5, pp. 613–625, 1993.

[63] A. Katsaggelos, J. Biemond, R. Schafer, and R. Mersereau, "A regularized iterative image restoration algorithm," *IEEE Transactions on Signal Processing*, vol. 39, no. 4, pp. 914–929, 1991.

[64] A. Tikhonov, A. Goncharsky, and V. Stepanov, "Inverse problem in image processing," in *Ill-Posed Problems in the Natural Sciences*, (Moscow), pp. 220–232, Mir Publishers, 1987.

[65] M. Sezan and A. Tekalp, "Survey of recent developments in digital image restoration," *Optical Engineering*, vol. 29, pp. 393–404, May 1990.

[66] S. Geman and D. Geman, "Stochastic relaxation, Gibbs distribution, and the Bayesian restoration of images," *IEEE Transactions on Pattern Analysis and Machine Intelligence*, vol. 6, pp. 721–741, Nov. 1984.

[67] G. Demoment, "Image reconstruction and restoration: Overview of common estimation structures and problems," *IEEE Transactions on Acoustics, Speech and Signal Processing*, vol. 37, pp. 2024–2073, Dec. 1989.

[68] M. Sondhi, "Image restoration: The removal of spatially invariant degradations," *Proceedings of the IEEE*, vol. 60, pp. 842–853, July 1972.

[69] H. Andrews, "Digital image restoration: A survey," *IEEE Computer*, vol. 7, pp. 36–45, May 1974.

[70] B. Hunt, "Digital image processing," *Proceedings of the IEEE*, vol. 63, pp. 36–45, Apr. 1974.

[71] B. Frieden, "Image enhancement and restoration," in *Picture Processing and Digital Filtering* (T. Huang, ed.), (New York), Springer-Verlag, 1975.

[72] F. Roddier, "The effects of atmospheric turbulence in optical astronomy," in *Progress in Optics* (E. Wolf, ed.), (Amsterdam, North-Holland), pp. 281–376, 1981.

[73] J. Oakley and B. Satherley, "Improving image quality in poor visibility conditions using a physical model for constrast degradation," *IEEE Transactions on Image Processing*, vol. 17, Feb. 1998.

[74] E. McCartney, *Optics of the Atmosphere: Scattering by Molecules and Particles*. New York: Wiley and Sons, 1976.

[75] V. Zuev, *Propagation of Visible and Infrared Radiation in the Atmosphere*. New York: Wiley and Sons, 1974.

[76] I. Dinstein, H. Zoabi, and N. Kopeika, "Prediction of effects of weather on image quality: Preliminary results of model validation," *Applied Optics*, vol. 27, pp. 2539–2545, 1998.

[77] S. Osher and L. I. Rudin, "Feature-oriented image enhancement using shock filters," *SIAM Journal on Numerical Analysis*, vol. 27, no. 4, pp. 919–940, 1990.

[78] P. Zemperoni, "Some adaptive rank order filters for image enhancement," *Pattern Recognition Letters*, vol. 11, no. 2, pp. 81–86, 1990.

[79] H. R. Rabiee, R. L. Kashyap, and S. R. Safavian, "Multiresolution segmentation-based image coding with hierarchical data structures," in *Proc. IEEE Int. Conf. on Acoustics, Speech and Signal Proc.*, pp. 1870–1873, 1996.

[80] L. Shen and M. M. Rangayyan, "A segmentation-based lossless image coding method for high-resolution medical image compression," *IEEE Trans. Medical Imaging*, vol. 16, no. 3, pp. 301–307, 1997.

[81] D. Wang, C. Labit, and J. Ronsin, "Segmentation-based motion-compensated video coding using morphological filters," *IEEE Trans. Circuits and Systems for Video Technology*, vol. 7, no. 3, pp. 549–555, 1997.

[82] D. Levi and S. Klein, "Weber law for position: The role of spatial frequency and contrast," *Vision Research*, vol. 32, pp. 2235–2250, 1992.

[83] R. Hess and I. Holliday, "The coding of spatial position by the human visual system: Spatial scale and contrast," *Vision Research*, vol. 32, pp. 1085–1097, 1992.

[84] G. Anderson and A. Netravali, "Image restoration based on a subjective criterion," *IEEE Transactions on Systems, Man and Cybernetics*, vol. 6, pp. 845–853, Dec. 1976.

[85] J. Beck, B. Hope, and A. Rosenfeld Eds., *Human and Machine Vision*. Orlando, FL: Academic Press, 1983.

[86] J. Beck, "Effect of orientation and shape similarity on perceptual grouping," *Peception and Psychophysics*, vol. 1, pp. 300–302, 1966.

[87] A. Gagalowicz, "A new method for texture field synthesis: Some applications to the study of human vision," *IEEE Transactions on Pattern Analysis and Machine Intelligence*, vol. 3, pp. 520–533, 1981.

[88] B. Julesz, "Experiments in the visual perception of texture," *Scientific American*, vol. 232, no. 4, pp. 34–43, 1975.

[89] M. Fahle and T. Poggio, "Visual hyperacuity: Spatiotemporal interpolation in human stereo vision," in *Proceedings of the Royal Society, London, UK*, vol. B213, pp. 451–477, 1981.

[90] D. Marr, S. Ullman, and T. Poggio, "Bandpass channels, zero-crossings and early visual information processing," *Journal of the Optical Society of America*, vol. 69, pp. 914–916, 1979.

[91] F. Attneave, "Some informational aspects of visual perception," *Psychological Review*, vol. 61, pp. 183–193, 1954.

[92] G. Johansson, "Visual motion perception," *Scientific American*, vol. 232, no. 6, pp. 76–88, 1975.

[93] R. Hess and D. Badcock, "Metric for separation discrimination by the human visual system," *Journal of the Optical Society of America*, vol. 12, pp. 3–16, Jan. 1995.

[94] T. Hentea and V. Algazi, "Perceptual models and the filtering of high contrast achromatic images," *IEEE Transactions on Systems, Man and Cybernetics*, vol. 14, pp. 230–246, Apr. 1984.

[95] V. Algazi, G. Ford, and H. Chen, "Linear filtering of images based on properties of vision," *IEEE Transactions on Image Processing*, vol. 4, pp. 1460–1464, Oct. 1995.

[96] X. Ran and N. Farvardin, "A perceptually motivated three-component image model. Part I: Description of the model," *IEEE Transactions on Image Processing*, vol. 4, pp. 401–415, Apr. 1995.

[97] A. Bellini, A. Leone, and R. Rovatti, "Fuzzy perceptual filtering for still image compression," in *Proceedings of the IEEE International Joint Conference on Neural Networks*, Anchorage, USA, May 5-9, pp. 1637–1642, 1998.

[98] I. Hontsch and L. Karam, "Locally-adaptive image coding based on a perceptual target distortion," in *Proceedings of the IEEE International Conference on Acoustics, Speech and Signal Processing*, Seattle, USA, May 12-15, vol. V, p. 2569, 1998.

[99] J. Huang and E. Coyle, "Perceptually optimal restoration of images with stack filters," in *Proceedings of the IEEE International Conference on Acoustics, Speech and Signal Processing*, Seattle, USA, May 12-15, vol. V, p. 2873, 1998.

[100] S. Daly, "The visible difference predictor: an algorithm for the assessment of image fidelity," in *Digital Images and Human Vision* (A. Watson, ed.), Cambridge, MA: MIT Press, 1993.

[101] S. Perry and L. Guan, "Weight assignment for adaptive image restoration by neural networks," *IEEE Transactions on Neural Networks*, vol. 11, pp. 156–170, Jan. 2000.

[102] S. Perry, *Adaptive Image Restoration: Perception Based Neural Network Models and Algorithms*. PhD thesis, School of Electrical and Information Engineering, University of Sydney, Apr. 1999.

[103] J. P. Sutton, J. S. Beis, and L. E. H. Trainor, "A hierarchical model of neuron-cortical synaptic organization," *Math. Comput. Modeling*, vol. 11, pp. 346–350, 1988.

[104] I. Pitas and A. Venetsanopoulos, *Nonlinear Digital Filters: Principles and Applications*. Boston: Kluwer Academic Publishers, 1990.

[105] J. K. Paik and A. K. Katsaggelos, "Image restoration using a modified Hopfield network," *IEEE Trans. Image Processing*, vol. 1, no. 1, pp. 49–63, 1992.

[106] Y. Iiguni *et al.*, "A real-time learning algorithm for a multilayered neural network based on the Extended Kalman filter," *IEEE Transactions on Signal Processing*, vol. 40, pp. 959–966, Apr. 1992.

[107] R. Steriti and M. A. Fiddy, "Regularised image reconstruction using SVD and a neural network method for matrix inversion," *IEEE Transactions on Signal Processing*, vol. 41, pp. 3074–3077, Oct. 1993.

[108] S. Perry and L. Guan, "Neural network restoration of images suffering space-variant distortion," *Electronic Letters*, vol. 31, pp. 1358–1359, Aug. 1995.

[109] S. Perry and L. Guan, "A partioned modified Hopfield neural network algorithm for real-time image restoration," *Journal of Real-Time Imaging*, vol. 2, pp. 215–224, 1996.

[110] S. Perry and L. Guan, "Restoration of images degraded by space variant distortion using a neural network," in *Proceedings of the IEEE Conference on Neural Networks*, Perth, Australia, Nov.26-Dec.2, vol. 4, pp. 2067–2070, 1995.

[111] S. Perry and L. Guan, "Image restoration using a neural network with an adaptive constraint factor," in *Proceedings of the IEEE Conference on Neural Networks in Signal Processing*, Nanjing, P.R. China, Dec. 10-20, pp. 1031–1034, 1995.

[112] S. Perry and L. Guan, "Image restoration using a neural network with an adaptive constraint factor (invited paper)," in *Proceedings of Computational Engineering in Systems Applications,* Lille, France, July. 9-12, pp. 854–859, 1996.

[113] M. G. Kang and A. K. Katsaggelos, "General choice of the regularization functional in regularized image restoration," *IEEE Trans. Image Processing,* vol. 4, no. 5, pp. 594–602, 1995.

[114] J. Anderson and J. Sutton, "A networks of networks: Computation and neurobiology," *World Congress of Neural Networks,* vol. 1, 1995.

[115] S. Perry and L. Guan, "A statistics-based weight assignment in a Hopfield neural network for adaptive image restoration," in *Proceedings of the IEEE International Joint Conference on Neural Networks,* Anchorage, USA, May 5-9, pp. 992–997, 1998.

[116] S. Perry and L. Guan, "Adaptive constraint restoration and error analysis using a neural network," in *Proceedings of Australian Joint Conference on Artificial Intelligence,* Perth, Australia, pp. 87–95, 1997.

[117] E. Chong and S. Zäk, *An Introduction to Optimization.* New York, USA: John Wiley and Sons, 1996.

[118] S. Perry and L. Guan, "Peception based adaptive image restoration," in *Proceedings of the IEEE International Conference on Acoustics, Speech and Signal Processing,* Seattle, USA, May 12-15, vol. V, pp. 2893–2896, 1998.

[119] B. L. M. Happel and J. M. J. Murre, "Design and evolution of modular neural network architectures," *Neural Networks,* vol. 7, no. 6-7, pp. 985–1004, 1994.

[120] R. A. Jacobs and M. I. Jordan, "Learning piecewise control strategies in a modular neural network architecture," *IEEE Trans. Syst. Man Cybern.,* vol. 23, no. 2, pp. 337–345, 1993.

[121] A. Kehagias and V. Petridis, "Time-series segmentation using predictive modular neural networks," *Neural Computation,* vol. 9, no. 8, pp. 1691–1709, 1997.

[122] S. Ozawa, K. Tsutsumi, and N. Baba, "An artificial modular neural network and its basic dynamical characteristics," *Biological Cybernetics,* vol. 78, no. 1, pp. 19–36, 1998.

[123] L. Wang, S. A. Rizvi, and N. M. Nasrabadi, "A modular neural network vector predictor for predictive image coding," *IEEE Trans. Image Proc.,* vol. 7, no. 8, pp. 1198–1217, 1998.

[124] L. Wang, S. Z. Der, and N. M. Nasrabadi, "Automatic target recognition using a feature-decomposition and data-decomposition modular neural network," *IEEE Trans. Image Proc.*, vol. 7, no. 8, pp. 1113–1121, 1998.

[125] H. Fu and Y. Y. Xu, "Multilinguistic handwritten character recognition by Bayesian decision-based neural networks," *IEEE Trans. Signal Proc.*, vol. 46, no. 10, pp. 2781–2789, 1998.

[126] S. Y. Kung, M. Fang, S. P. Liou, M. Y. Chiu, and J. S. Taur, "Decision-based neural network for face recognition system," in *Proc. Int. Conf. on Image Processing*, pp. 430–433, 1995.

[127] S. H. Lin and S. Y. Kung, "Probabilistic DBNN via expectation-maximization with multi-sensor classification applications," in *Proc. Int. Conf. on Image Processing*, pp. 236–239, 1995.

[128] S. H. Lin, S. Y. Kung, and L. J. Lin, "Face recognition/detection by probabilistic decision-based neural networks," *IEEE Trans. Neural Networks*, vol. 8, no. 1, pp. 114–132, 1997.

[129] A. C. Bovik, T. S. Huang, and D. C. Munson, "A generalization of median filtering using linear combinations of order statistics," *IEEE Trans. Acoust., Speech, Signal Processing*, vol. 31, no. 6, pp. 1342–1350, 1983.

[130] M. Ropert, F. M. de Saint-Martin, and D. Pele, "A new representation of weighted order statistic filters," *Signal Processing*, vol. 54, no. 2, pp. 201–206, 1996.

[131] J. T. Tou and R. C. Gonzalez, *Pattern Recognition Principles*. Reading, MA: Addison-Wesley, 1974.

[132] L. A. Zadeh, "Fuzzy algorithms," *Information and Control*, vol. 12, no. 2, pp. 94–102, 1968.

[133] L. A. Zadeh, "Outline of a new approach to the analysis of complex systems and decision processes," *IEEE Trans. Syst. Man Cybern.*, vol. 3, no. 1, pp. 28–44, 73.

[134] L. A. Zadeh, "Fuzzy sets as a basis for a theory of possibility," *Information and Control*, vol. 1, no. 1, pp. 3–28, 1978.

[135] J. Hopfield and D. Tank, "Neural computation of decisions in optimization problems," *Biol. Cybern.*, vol. 52, no. 3, pp. 141–152, 1985.

[136] L. J. Fogel, A. J. Owens, and M. J. Walsh, *Artificial Intelligence Through Simulated Evolution*. New York: Wiley, 1966.

[137] T. Bäck, D. B. Fogel, and Z. Michalewicz, eds., *Handbook of Evolutionary Computation*. New York: Oxford University Press and Institute of Physics, 1997.

[138] T. Bäck, U. Hammel, and H.-P. Schwefel, "Evolutionary computation: Comments on the history and current state," *IEEE Trans. Evolutionary Comp.*, vol. 1, no. 1, pp. 3–17, 1997.

[139] T. Bäck and H.-P. Schwefel, "Evolutionary computation: An overview," in *Proc. 3rd IEEE Conf. on Evolutionary Computation*, (Piscataway, NJ), pp. 20–29, IEEE Press, 1996.

[140] J. H. Holland, *Adaptation in Natural and Artificial Systems*. Ann Arbor, MI: University of Michigan Press, 1975.

[141] M. Mitchell, *An Introduction to Genetic Algorithms*. Cambridge MA: MIT Press, 1996.

[142] H.-P. Schwefel, *Numerical Optimization of Computer Models*. Chichester: Wiley, 1981.

[143] H.-P. Schwefel, *Evolution and Optimum Seeking*. New York: Wiley, 1995.

[144] K. H. Yap and L. Guan, "Adaptive image restoration based on hierarchical neural networks," *Optical Engineering*, vol. 39, no. 7, pp. 1877–1890, 2000.

[145] W. Pratt, *Digital Image Processing*. New York: Wiley and Sons, 1991.

[146] M. Cannon, "Blind deconvolution of spatially invariant image blur with phase," *IEEE Trans. Acoust. Speech, Signal Processing*, vol. 24, pp. 58–63, 1976.

[147] D. Kundur and D. Hatzinakos, "Blind image deconvolution," *IEEE Signal Processing Magazine*, vol. 13, no. 3, pp. 43–64, 1996.

[148] R. L. Lagendijk, J. Biemond, and D. E. Boekee, "Identification and restoration of noisy blurred images using the expectation-maximization algorithm," *IEEE Trans. Acoust. Speech, Signal Processing*, vol. 38, pp. 1180–1191, July 1990.

[149] S. Reeves and R. Mersereau, "Blur identification by the method of generalized cross-validation," *IEEE Transactions on Image Processing*, vol. 1, pp. 301–311, July 1992.

[150] G. Ayers and J. Dainty, "Iterative blind deconvolution method and its applications," *Optical Letters*, vol. 13, pp. 547–549, 1988.

[151] B. C. McCallum, "Blind deconvolution by simulated annealing," *Optics Communications*, vol. 75, pp. 101–105, Feb. 1990.

[152] D. Kundur and D. Hatzinakos, "A novel blind image deconvolution scheme for image restoration using recursive filtering," *IEEE Trans. Signal Processing*, vol. 46, pp. 375–390, Feb. 1998.

[153] Y. You and M. Kaveh, "A regularization approach to joint blur identification and image restoration," *IEEE Transactions on Image Processing*, vol. 5, pp. 416–428, Mar. 1996.

[154] T. F. Chan and C.-K. Wong, "Total variation blind deconvolution," *IEEE Trans. Image Processing*, vol. 7, pp. 370–375, Mar. 1998.

[155] K.-H. Yap and L. Guan, "A recursive soft-decision PSF and neural network approach to adaptive blind image regularization," in *Proc. IEEE Int. Conf. Image Processing*, vol. 3, pp. 813–816, Sept. 2000.

[156] D. B. Fogel, "An introduction to simulated evolutionary optimization," *IEEE Trans. Neural Networks*, vol. 5, pp. 3–14, Jan. 1994.

[157] K. S. Tang, K. F. Man, S. Kwong, and Q. He, "Genetic algorithms and their applications," *IEEE Signal Processing Magazine*, pp. 22–37, Nov. 1996.

[158] J. P. Sutton, *Hierarchical Organization and Disordered Neural Systems*. PhD thesis, University of Toronto, 1988.

[159] L. Guan, J. Anderson, and J. Sutton, "A network of networks model for image regularization," *IEEE Trans. on Neural Networks*, vol. 8, pp. 169–174, Jan. 1997.

[160] A. A. Zhigljavsky, *Theory of Global Random Search. Mathematics and Its Applications*. Dordrecht: Kluwer, 1992.

[161] P. Baldi and K. Hornik, "Neural networks and principal component analysis: learning from examples without local minima," *Neural Networks*, vol. 2, no. 1, pp. 53–58, 1989.

[162] E. Oja, "A simplified neuron model as a principal component analyzer," *Journal of Mathematical Biology*, vol. 15, no. 3, pp. 267–273, 1982.

[163] E. Oja, "Neural networks, principal components and subspaces," *Int. J. Neural Systems*, vol. 1, no. 1, pp. 61–68, 1989.

Index